Advanced engineering design

Related titles:

Aerodynamic measurements: From physical principles to turnkey instrumentation
(ISBN 978-1-84569-992-5)
Aerodynamic measurements will present to its reader a comprehensive treatment of the theoretical bases which experimental techniques used in aerodynamics are based on. Limitations of each method in terms of accuracy, response time and complexity will also be addressed. This book will serve as a guide in choosing the most pertinent technique for each type of flow field including: 1D, 2D, 3D, steady or unsteady, subsonic, supersonic or hypersonic. This is currently no book that presents as many techniques as those presented in this volume, with such a critical review of the various methods offered. *Aerodynamic measurements* will help guide the reader to choose the most appropriate in each case and for each technique the evolution from old-fashioned mechanical process to the modern computerized version is reported in order to allow students and practitioners to understand what is behind the results shown on the coloured screen of their computers.

Spacecraft thermal control
(ISBN 978-1-84569-996-3)
This book covers the necessary material to allow an engineer to perform the design and qualification of a Space Thermal Control System in a comprehensive way. Beginning with the fundamentals of Space Missions and following with the indispensable background on Heat Transfer, a full description of current thermal control technologies is then presented. The overall view presented in *Spacecraft thermal control* details the complete process in the design and validation (analysis and tests) of these systems. Finally relevant data on material properties are given.

Orbital mechanics and formation flying: A digital control perspective
(ISBN 978-0-85709-054-6)
Aimed at students, faculty and professionals in the aerospace field, this book provides practical information in the development, analysis, and control of a single and/or multiple spacecraft in Space. This book is divided into two major sections: single and multiple satellite motion. The first section will analyze the orbital mechanics, orbital perturbations, and attitude dynamics of a single satellite about the Earth. Using the knowledge of a single satellite motion, the translation of a group of satellites called formation flying or constellation is explained. Formation flying is one of the main research topics in the last few years and this book explains different control approaches to control the satellite attitude motion and/or to maintain the constellation together. The control schemes are explained in the discrete domain such that it can be easily implemented on the computer onboard the satellite. The key objective of the book is to show the reader the practical and the implementation process in the discrete domain.

Advanced engineering design

An integrated approach

Efrén Moreno Benavides

WP
WOODHEAD PUBLISHING

Oxford Cambridge Philadelphia New Delhi

Published by Woodhead Publishing Limited, 80 High Street, Sawston, Cambridge CB22 3HJ, UK
www.woodheadpublishing.com

Woodhead Publishing, 1518 Walnut Street, Suite 1100, Philadelphia, PA 19102–3406, USA

Woodhead Publishing India Private Limited, G-2, Vardaan House, 7/28 Ansari Road, Daryaganj, New Delhi – 110002, India
www.woodheadpublishingindia.com

First published 2012, Woodhead Publishing Limited

British Library Cataloguing in Publication Data
A catalogue record for this book is available from the British Library.

Library of Congress Control Number: 2011942110

Woodhead Publishing ISBN 978-0-08-101663-3 (print)
ISBN 978-0-85709-504-6 (online)

Typeset by RefineCatch Limited, Bungay, Suffolk

Dedicated to Carmela, Elías and Celia.

Contents

List of figures and tables

Figures

Tables

Acknowledgments

I would like to express my gratitude . . .

For the enriching research of Nam Pyo Shu and Genichi Taguchi. Their works are, without a doubt, the fundamental pillar on which I built the discourse presented in this book. Their contributions give meaning to the concept of 'best' in engineering, and serve as a unifying thread through the different chapters. Thanks to them, I took on the challenge of seeking objectivity in a field that traditionally suffers from a lack of it: the definition of the best engineering design process.

To all the professors at the School of Aeronautical Engineering at the Technical University of Madrid, and most especially those who had me as a student. Thanks to them, I received the necessary training to seek precision and rigor in the reflections on which engineering is based. I am also very grateful to all of my students because, thanks to their questions, I have been forced to find better answers. And thanks to their continuous encouragement, I have found the necessary strength to continue this search even when other tasks might have seemed more pressing.

To Alfonso Martínez Fernández and Fernando de la Iglesia Moreno for trusting in my proposals and giving me encouragement, good advice, and support from ALTRAN Technologies. I am doubly grateful because they managed to redirect my ideas without destroying them, while also having the patience to watch them grow.

To Massimo Bandecchi for providing me with access to the charts produced through his work in concurrent engineering, taken from the European Space Agency.

To David Howie, Mike Wood and Beatriz Sánchez Barba for giving me access to the illustration of the beautiful Trent 500 engine by Rolls-Royce.

To Juan Manuel Tizón Pulido, Joan Bruno Rodríguez Pastor and Luis García Rodríguez because their comments, reflections, questions and advice helped me to polish the way some ideas are presented in the book.

To Tania Michelle Tate for translating the text, and accompanying me as the original Spanish manuscript was transformed into the final version that is this book; and also for her patience when some of the finished chapters underwent drastic modifications.

Finally, to the editors and other staff at Chandos Publishing House for their magnificent work.

Efrén M. Benavides
1 July 2011

About the author

Efrén M. Benavides is a Professor at the *Escuela de Ingeniería Aeronáutica y del Espacio* (EIAE), the largest School of Aeronautical and Space Engineering in Spain. He currently works in the Propulsion and Fluid Mechanics Department, which belongs to the Polytechnic University of Madrid (UPM). He teaches courses in 'Turbocharging of Reciprocating Engines' (final-year project), 'Mechanical Design' (fifth year), 'Reciprocating Engines' (fourth and fifth years), 'Propulsion Systems' (second year) for undergraduate Aeronautical Engineering students and 'Introduction to Propulsion with Piston Engines' for Master's students in Test Flight and Aircraft Certification. He is responsible for the 'Sustainable Design Theories' course for Master's students in Environmental Sciences, and was in charge of the Postgraduate course 'Mechanical Design with Total Quality; Design of Experiments' for Ph.D. students.

His academic background includes a six-year degree in Aeronautical Engineering (1994, Polytechnic University of Madrid), a Ph.D. in High-Pressure Fuel Flow Simulation (1999) and a five-year degree in Physics (2005, Complutense University of Madrid). Following an initial period of collaboration with Spain's National Institute for Aerospace Technology (INTA) in the design, manufacture and testing of aluminum high-pressure fuel pumps, he began his career as a researcher. As a result, Efrén is the author of several JCR scientific articles covering different aspects of Engineering: thermoacoustic engines; non-stationary heat transfer; stability of axial-flow compressors; and reliability of devices subjected to variable stresses. He is also the author of two book chapters on heat transfer and diesel engines ('Heat Transfer Enhancement for Weakly Oscillating Flows' in *Heat Transfer – Theoretical analysis, experimental investigations and industrial systems* and 'Numerical Simulation of the Injection Process in a Two-Stroke Diesel Engine' in *SAE 2000 Transactions: Journal of Engines*), as well as a philosophical essay (the book *Ética Borrosa*, published in Spanish). In addition, he has combined research with industry by collaborating with a major consulting firm, ALTRAN Technologies, to develop a methodology that implements Axiomatic Design as a part of its consulting strategy.

The author may be contacted at:
Propulsion and Fluid Mechanics Department
ETSI Aeronáuticos – UPM
EIAE – UPM
Pza. Cardenal Cisneros, 3
28040 Madrid
Spain
E-mail: efren.moreno@upm.es

1

The design process

Abstract: Whether a design is a work of art, intended to provoke a reflection or succession of emotions in the viewer, or an engineering system, intended to meet a technological need, there is a motivation. The existence of this motivation is the fundamental premise on which the ideas set forth in this book are based. The purpose of this chapter is to explore the most basic concepts covered by this statement, and the objective of the later chapters is to develop these concepts in greater depth.

Key words: motivation, need, requirements, information, available information, efficiency, robustness, mediator, designer, producer, evaluator, customer, supplier, value, iteration, feedback, analysis, synthesis, conceptual design, detailed design, concurrent design.

1.1 The concept of design and related terms

Let us accept as a premise that all design is a response to a particular motivation. Design and motivation are therefore concepts mutually linked by a causal relationship: the former is the consequence of the latter, and the latter is necessary for the former to exist. Thus, to understand a design, it is necessary to understand its motivation. The motivation is the cause or reason that moves something. In this context, the motivation moves the designer to design, and the purpose of the design is to respond to the motivation.

However, the existence of a motivation is not a sufficient condition for the existence of a satisfactory design. A design process must necessarily take place between the motivation and the conclusion of the design, which can make it impossible, difficult, or easy for the design to be achieved. For example, if the design process requires more resources than are available, it will not be completed. The designer in charge of responding to the motivation always consumes resources in order to generate the response. At the very least, the time spent searching for an approach cannot be devoted to other activities. It is therefore appropriate to rewrite the initial premise so that the existence of a resource-consuming process is explicitly reflected in its formulation: *design is the result of consuming resources in order to provide a response to a motivation*. The relationship between these basic ingredients is shown in schematic form in Figure 1.1.

Defining a *motivation* requires finding the motive, cause, reason or purpose for it. To define the motivation is to reformulate it so that all the indispensable characteristics needed to communicate it appear explicitly in the wording. Because they are all indispensable, if any of these characteristics were eliminated from the definition, the definition would change, and with it the motivation. The complete list of characteristics

Figure 1.1 **Design as a consequence of the resources consumed by a motivation**

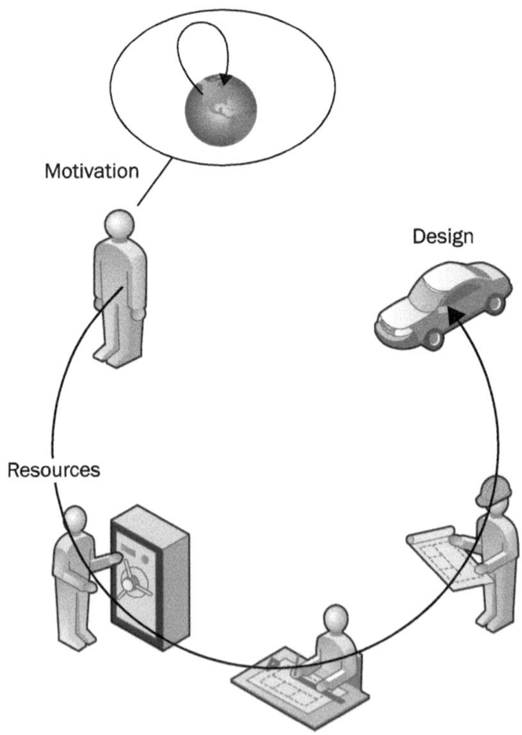

that define the motivation is therefore necessary to identify it. Finding a way to meet the characteristics specified on the list, in response to the motivation, is also a need. For this reason, we will refer to the set of necessary characteristics as a *need* or *list of needs*. Because the list of needs defines the motivation, it can be used as a substitute for it in the definition of the problem. The fundamental difference between the motivation and the list of needs is the information available to properly define the motivation. When the *available information*[1] is adequate, the motivation and the list of needs (the need), can be considered two equivalent formulations of the same problem.[2] However, when the available information is not complete, or adequate, the same motivation can generate different lists of needs; in other words, different formulations of the problem. This relationship is shown in Figure 1.2.

In the context of engineering, it is common to define a design problem, or motivation, in terms of *requirements*. However, in this text we will refer to 'the list of needs,' 'needs,' or 'need' when no condition is imposed on them other than describing the motivation. This convention enables us to reserve the word 'requirements' for expressing the definition provided by Nam P. Suh in the context of Axiomatic Design.[3]

Finding the response that satisfies the motivation with the lowest possible resource consumption, i.e. the most efficient design process,

Figure 1.2 **The relationship between needs, motivation and information**

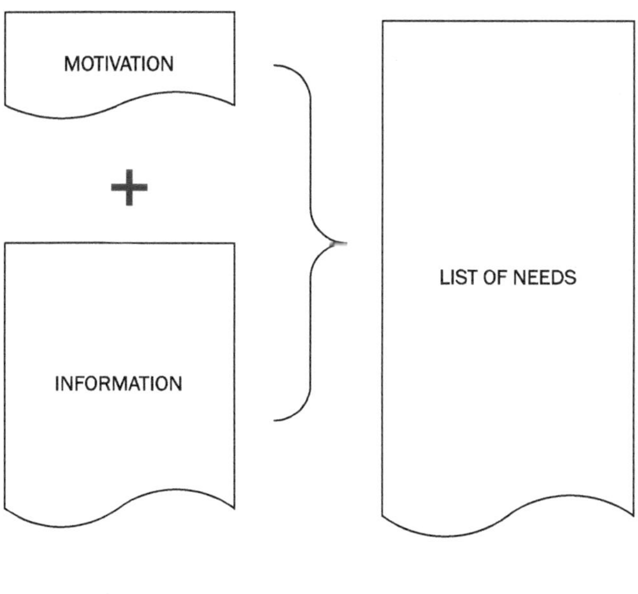

means reducing the number of possible designs to be studied. From this perspective, *efficiency* is a restriction for both the design process itself and the final product designed. However, from another perspective, if we accept that responding to motivations inevitably consumes resources, which are inevitably limited, we also accept that efficiency[4] is a need. This is how the list of needs defines and delimits the motivation. To summarize, the following is a list of terms related to the design process.

Terms relating to the *approach to the problem*:

- **Motivation.** Cause, reason, or purpose that moves something.
- **Need or list of needs.** Those circumstances, conditions, characteristics, qualities, or demands that may not be lacking without distorting the description of the motivation.
- **List of requirements.** In the context of Axiomatic Design, circumstances or conditions necessary for describing the motivation. This is different from the list of needs because it has additional limitations: they must comprise a minimum, independent list.
- **Constraints.** Those needs that may not be included in the list of requirements.

Terms relating to the *approach to the response*:

- **Satisfy.** To provide a solution to a difficulty arising from a motivation by meeting the needs, requirements and constraints.
- **Solve.** To provide a solution to a difficulty arising from a motivation, considering the needs, requirements and constraints. The difference with regard to 'satisfy' is that although the needs, requirements and constraints are taken into account, compliance is not assured.

1.2 Design as a value-generating process in society

Those members of society who have needs are the customers, while those who ensure that the needs are satisfied are the suppliers. In the context of this book, we will refer to the people sharing a particular need as the *customers*, although they may be a group and might only be potential customers. (Their need may be a current reality, or it might be latent and not yet visible.) For example, they might feel that some internal need is satisfied while admiring a work of art, to the point that they wish to purchase it. Given the general nature of the definition used here, it obviously applies to works of engineering as well as works of art. In the context of this book, the people who share the task of satisfying a customer are referred to as the *suppliers*. A supplier can in turn become the customer of another supplier when satisfying the motivation of the first customer becomes a new motivation associated with a list of needs. The systematic repetition of this customer–supplier pattern comprises the hierarchy of needs. Figure 1.3 shows the customer–supplier relationship, in which the customer's need is satisfied by a supplier by consuming resources.

Normally, the supplier must perform a wide range of tasks in order to achieve his purpose, but the initial premise that any design is a response to a particular motivation forces the supplier to clearly understand the motivation, and also forces the customer to clearly understand why the supplier's response satisfies his need. One of the functions assumed by the supplier is to mediate so that the customer clearly expresses his motivation and clearly understands the response. In addition to his activities as a mediator, the supplier must be capable of generating the adequate response to the customer's motivation. In order to do so, he also assumes the functions for designing a solution. When the supplier appoints managers for these functions, the pattern of customer–supplier relationships shown in Figure 1.3 is transformed, resulting in a situation similar to the one shown in Figure 1.4. Thus, the *mediator*, responsible for understanding the customer's motivation and adequately transmitting it, rephrases the motivation in such a way that the designer clearly

Figure 1.3 The value-generating process in society

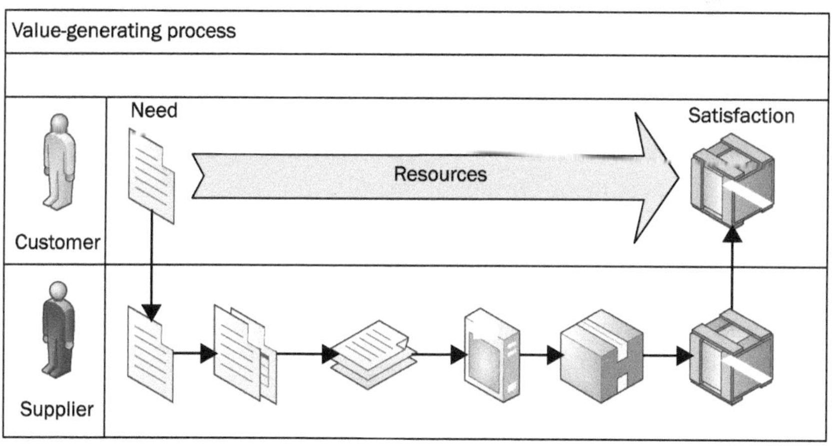

Figure 1.4 The role of the mediator in the value-generating process in society

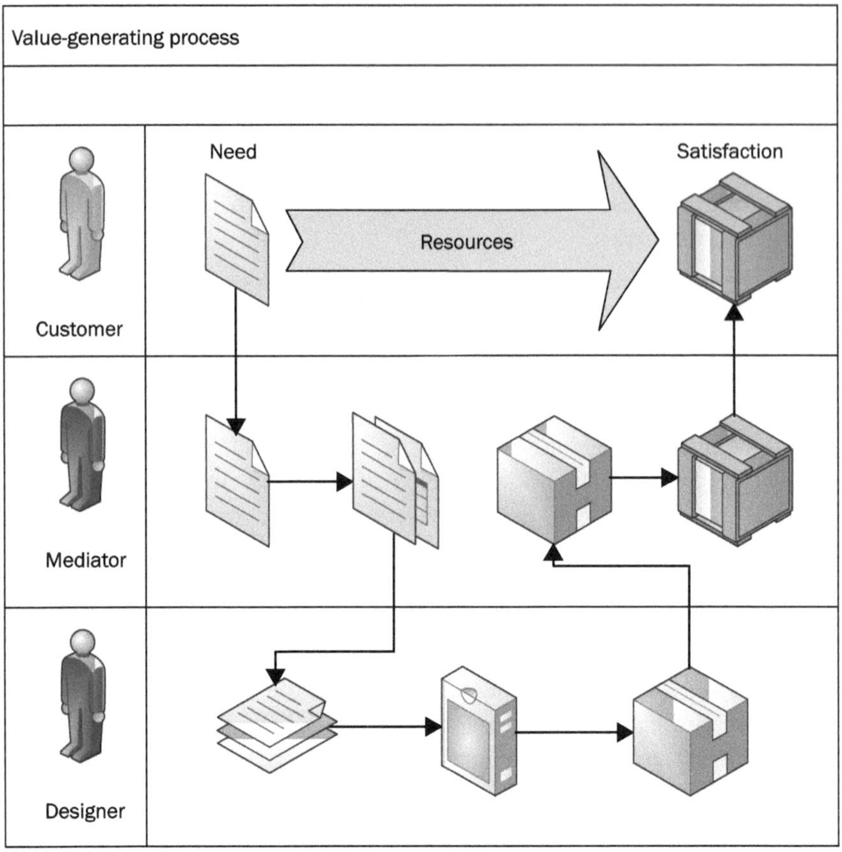

understands the need that must be satisfied. The mediator also puts the customer in touch with the appropriate designer. Once the designer offers a solution, the mediator will rephrase it in terms of the initial motivation to make it understood to the customer.

There is a person or group of people, the *designer* or group of designers, who create a product where nothing existed before: an emotive canvas in the case of an artist, or a spacecraft in the case of an engineer. In both cases, the designer must carry out a creative process in order to produce a product or solution that responds to the motivation and satisfies the need. Thus, one of the functions of the designer is to discover possible solutions: his most highly valued characteristic is *creativity*, and being a good creator is his raison d'être.

However, to ensure success, one of his functions must be to think of the possible solutions in such a way that they do not violate the restrictions imposed by the branch of knowledge in which he works. It would be useless to propose as a solution a vehicle that violates the laws of thermodynamics, or an electrostatic device that violates Earnshaw's theorem. Consequently, an important part of his job is to know the rules[5] of his discipline very well. In the case of engineering, he must find adequate, feasible solutions to the motivation. These solutions appear as results after a series of resources are consumed, e.g. the time and energy employed, and quite possibly after transforming another series of resources, such as the materials. All the operators responsible for appropriately transforming the materials for the purpose at hand must be part of the supplier, and are referred to as the *producers*.

The supplier also needs to employ certain *evaluators*. This is because the creative process is the result of the will and capacities of the people behind it, whose approach is therefore subjective and, to some extent, random. Thus, a multitude of designs can satisfy the same need, or respond to the same purpose. The evaluator is responsible for selecting what he believes to be the most suitable of all the possible designs. The relative arbitrariness underlying the creative process means that the number of possible designs is infinite. However, even so, there is something objective underlying each one, because it is well known that not all of them will be equally accepted. This gives rise to a third group of functions associated with the designer: those of evaluation. Obviously, the evaluator exists because there is a decision criterion that allows him to classify the responses, at least as good or bad, better or worse.[6] Managing these criteria is precisely the need met by the evaluator.

Thanks to this approach, we can classify the different actors intervening in the process by their functions. Figure 1.5 shows the functions associated with the supplier. Obviously, the three groups–customers, mediators and designers–might not be disjoined. This often occurs even to the extreme, and not at all exceptional, of having a single common element: a person who designs a product for his own personal use is the customer, mediator, creator, producer and evaluator at the same time. The reality is much more complex, as one customer's initial need leads to the satisfaction of a set of secondary needs for other customers. These secondary customers, sometimes referred to as internal customers, are at least all members of the company or, in general, everyone involved in the process intended to satisfy the initial need. Among them, there will obviously be members belonging to the groups of creators, producers and evaluators. The

Figure 1.5 Functions associated with the different actors intervening to satisfy the customer's motivation

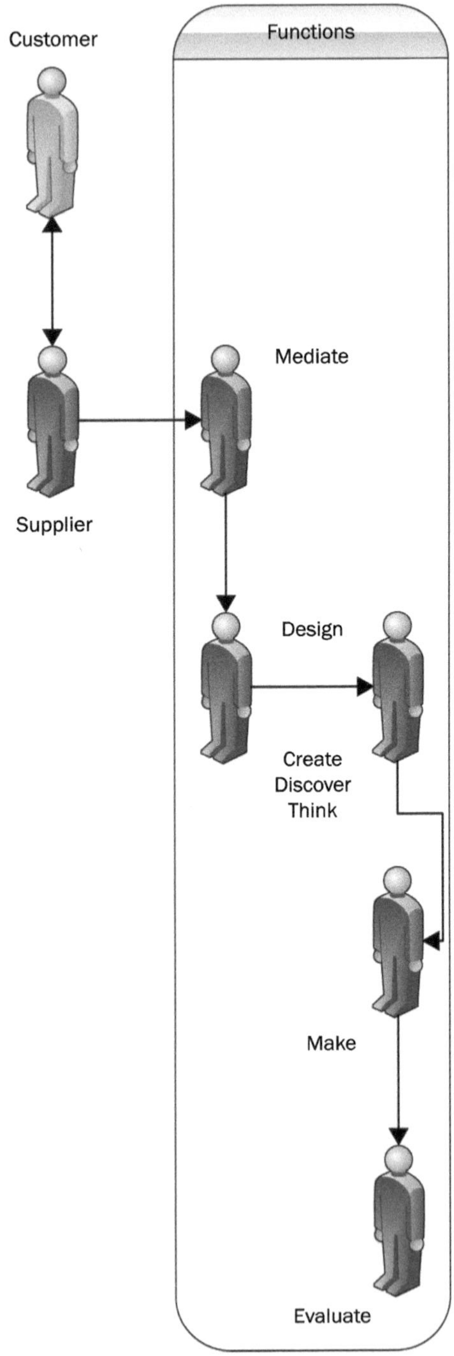

satisfaction of their need is the motivation driving them to take part in the process. This situation makes them all customers, some as direct consumers of the product, and others as consumers of other products as a result of their involvement with the initial one. The former pay with their money and receive the product, while the latter pay with their time and receive the money. In this way, the process exists and the primary need is satisfied. For this reason, the needs of everyone involved in the need–satisfaction cycle are as important as the initial need. All of these needs are indispensable in order for the product to exist.

Evaluators face the difficult problem of objectively classifying solutions as better or worse. Because the only elements unrelated to the subjectivity of the creator are the customer and the resources,[7] it is plausible to think that evaluators will base the indicator differentiating one design from another on the extent to which the customer's need has been satisfied and the quantity of resources consumed. In the context of this book, we will define this indicator as the value of a solution. In other words, we will classify solutions based on their value. Because value is generated in a society when a supplier satisfies a customer's need, we will define the *value* as the measure of satisfaction of that customer per unit of resources consumed.[8]

$$Value = \frac{Degree\ of\ satisfaction}{Resources\ consumed} \tag{1.1}$$

In the context of this introduction, the appropriate design will be the one that maximizes the value of the operation.[9] Naturally, the mediator must reinterpret the need and recreate the solution in order to maximize its value. Thanks to this scale, there is a substantial difference between each of the designs within the same branch of knowledge: it is obvious that not all designs will generate the same value, even if they all obey the rules imposed by the discipline in which they are developed and satisfy the customer equally. Figure 1.6 shows the value-generating process, taking into account the actors responsible for assuming the functions necessary for the design.

Figure 1.6 summarizes the ideas explained earlier, and clearly shows the drastic effects of the role of the evaluator on the process. To understand this, let's follow the steps in the value-generating process from the initial motivation in the form of a need until it is finally satisfied. First of all, the mediator transmits the need to the people responsible for suggesting possible responses to it. The role of the mediator is crucial from the moment he takes responsibility for defining the wording of the problem for the designers. The designers, taking into account the restrictions

Figure 1.6 **Actors intervening in the value-generating process in society**

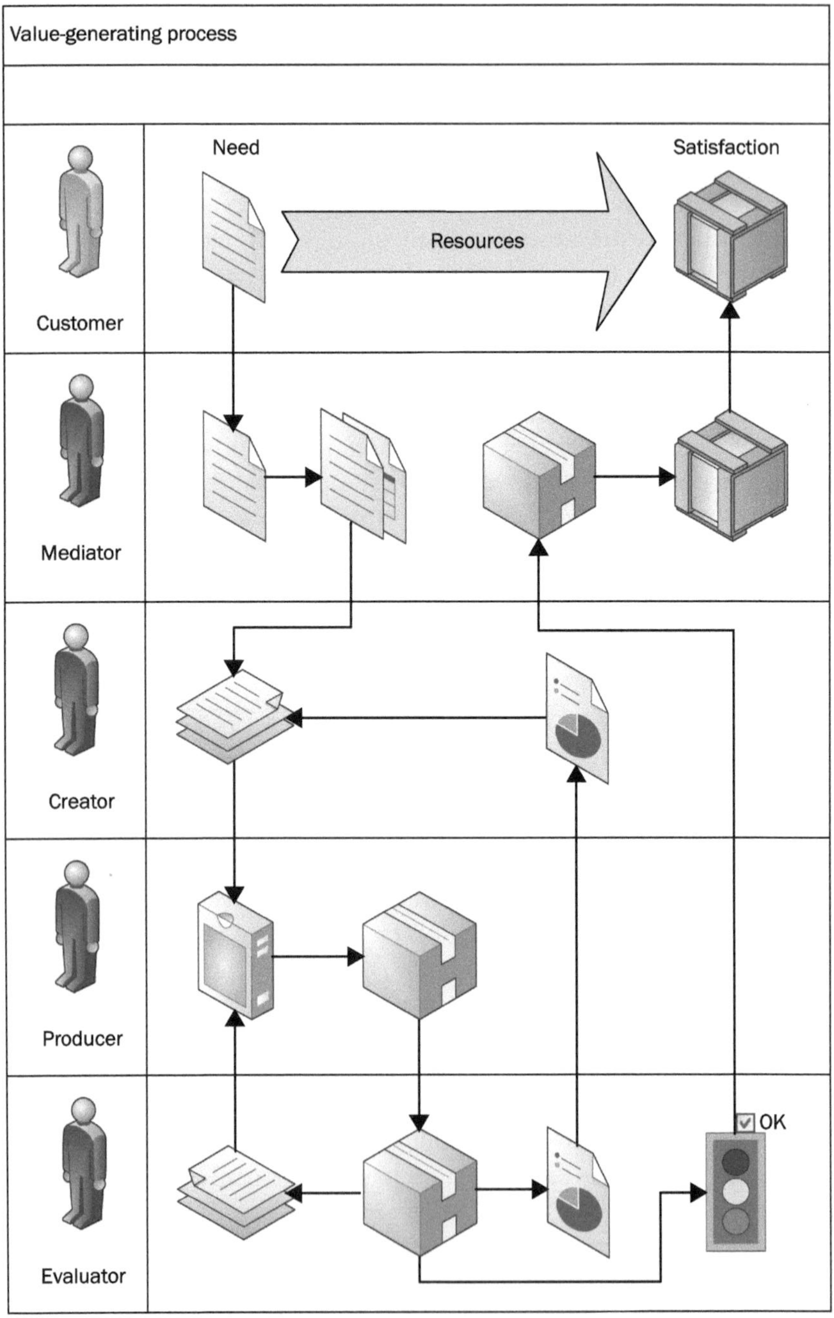

imposed by their discipline, suggest a solution that will be created and evaluated. The work of the evaluator leads to three types of actions. The first, identified with a stoplight in Figure 1.6, prevents the solution from being delivered to the customer as long as the value generated does not exceed a predefined minimum limit. The other two actions serve to inform the designers, during the 'thinking' and 'doing' phases, of the reasons why the product's value is not adequate. This information is added to the initial specifications to generate a new definition of the need and, consequently, the solution. This feedback process is repeated continuously in order to increase the value of the solution.

Each *iteration*, prompted by the *feedback* received from the evaluators in order to increase customer satisfaction, consumes resources. Thus, the value of the solution can increase if the satisfaction grows more than the resource consumption, decrease if it grows less, or remain constant. Therefore, only extraordinarily efficient design processes will maximize the value of the solutions. In poorly planned processes, the value of the solution can systematically decrease with each new attempt to improve upon it.

On the other hand, if the evaluators do not have a clear objective and repetitive procedure capable of evaluating customer satisfaction, the improvement process cannot be effective. It is also impossible to create a good product if the mediator does not initially reflect the customer's needs without distorting the real motivation. Although a correct approach to all of these activities is essential for maximizing the value of a solution, the tasks that are most repeated will have the greatest effect on the final value. Special care must therefore be taken with all activities involved in the two iterative processes shown in Figure 1.6, where the later stages of design, associated with the evaluation functions, provide feedback for the previous stages. These two key processes are shown in isolation in Figure 1.7.

The above reflections can be summarized as follows. The design process consumes resources and satisfies needs. In addition, the product generated by a design process is the result of human activity and, as such, has unique characteristics. For example, it is the result of a creative process oriented towards a purpose. This creativity is managed by the design process, which is part of the overall life cycle of the product and, as such, part of the product itself. Thus, finding a design process that generates well-evaluated designs is a need, and the correct design of the design process itself is therefore a purpose. This approach also leads us to describe the design process as a product subject to evaluation. It also becomes a universal objective from the moment it appears as

Figure 1.7 Two examples of possible feedback in a design process

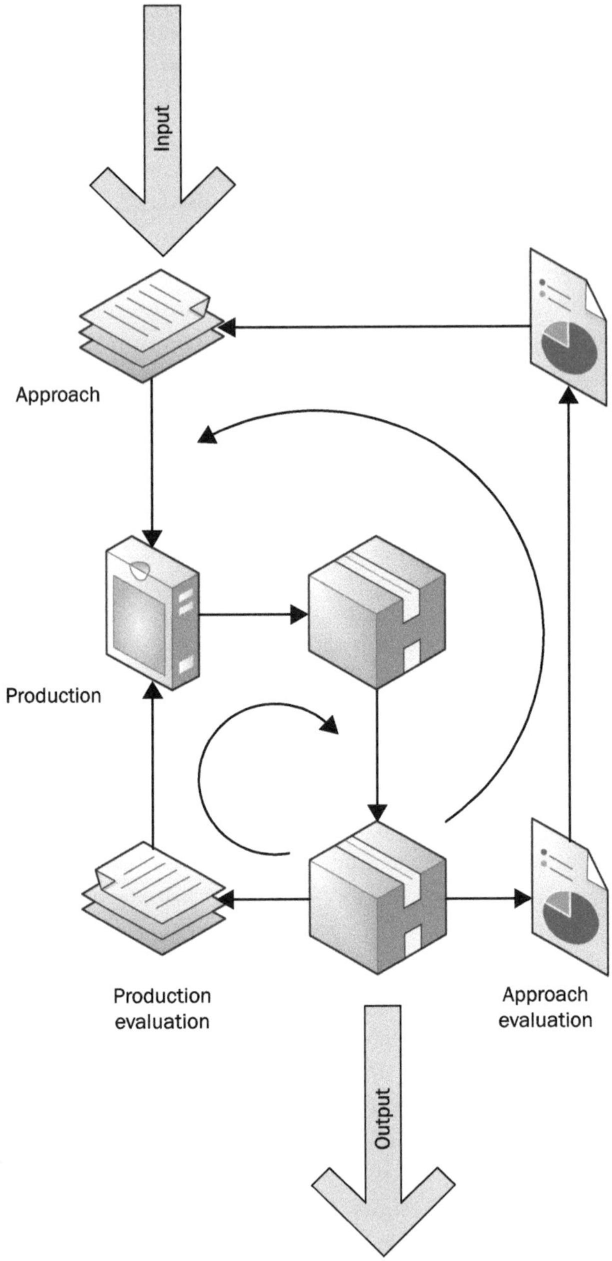

the engine of all human activities aimed at creating works subject to evaluation. This universality is what allows us to speak of Design Science.[10]

Knowledge of the basic rules of this science, in addition to the rules of one's own discipline, is indispensable for creating works that produce maximum satisfaction with minimum resource consumption. As we will see throughout this book, Design Science generates value to a much greater extent than other disciplines that are currently receiving more attention.

1.3 The goal of design theories

Design is creativity, consumes resources, has a purpose, and can therefore be assessed and evaluated. These are the four ingredients that enable us to build rules to guide creators in their work. However, this approach raises a concern. Is the set of rules that enable us to create an optimum design universal, objective and stable? If the answer is yes, we are dealing with a science. Let us therefore explore this abstraction process.

Consider the following motivation: to win the Formula 1 1968 French Grand Prix. We could, in turn, consider that this motivation means satisfying a need defined by the following list of needs: maximum power, minimum consumption, maximum stability, maximum reliability, maximum speed, maximum acceleration, maximum deceleration, etc. Given this difficult problem, there is a multitude of solutions. For example, engineers from Honda adopted the solution shown in Figure 1.8: the RA302. This car was regarded as one of the most innovative of its time, and for good reason. The RA302 had a light alloy chassis and a magnesium-skinned monocoque, on which they installed wishbone-type double suspension, disk brakes, a regular 5-speed transmission and an air-cooled 8-cylinder engine distributed into two 4-cylinder blocks forming a 120° V. It had a total cylinder capacity of 2987 cc, a compression ratio of 11.5 and 4 valves per cylinder with independent intake manifolds open directly to the atmosphere, transistorized electronic ignition, and low-pressure fuel injection. It was equipped with a fuel tank with a capacity of 200 liters, and the final weight of the vehicle was under 500 kg. Its maximum power was roughly 335 kW at 9500 rpm.

In spite of the impressive technological developments achieved by the team, the Honda RA302 did not respond to the motivation. The result

Figure 1.8 The Honda RA302 was the new solution adopted by the Honda F1 Racing Team to compete in the 1968 Formula 1 French Grand Prix. It competed along with the previous solution, the RA301

was devastating because the driver, Joseph Schlesser, lost his life. The accident occurred in the second lap of the Grand Prix, when the car ran off the track and crashed fatally. Analyzing the reasons behind this result or, in other words, why the motivation that generated that solution was not satisfied, is an extraordinarily complex problem involving engineering, human and environmental factors. Conjectures can be made regarding those factors unrelated to engineering that may have affected the outcome. Of these, two are difficult to ignore: John Surtees'[11] refusal to drive the one-seater, and the rain that was present on the circuit. As far as engineering factors are concerned, there were three major innovations: the use of a magnesium-skinned monocoque; the use of an air-cooled engine; and the change in the driver's position, which was brought forward in order to move the engine, and with it the vehicle's center of mass. This combination of factors proved to be fatal. Air-cooling, without the help of coolant, was maintained using dynamic intakes and a set of fins on the bottom. This tended to increase the engine temperature and reduce its life expectancy. In addition, not all of the cylinders received the same cooling, which resulted in temperature differences between them. Such a variable temperature field constituted a serious technological problem that the engineers had to solve by manufacturing parts with different tolerances in order to absorb the different thermal dilations. They also had to design the lateral air intakes to be capable of ingesting more air by taking advantage of the Venturi effect. All of these difficulties delayed and complicated the fine-tuning of the car. However, the placement of the engine in the center allowed for better mass distribution and, consequently, improved the dynamic behavior of the car. On the other hand, however, the position of the driver's legs so far forward over the front axis increased the possibility of him being seriously injured or trapped in the event of an accident, and caused the car to handle differently. (Figure 1.9 compares the position of the driver in the RA302 and the RA301, the other Honda one-seater participating in that Grand Prix.) Both effects may have played a decisive role in the tragic outcome. The first prevented the driver from exiting rapidly, and the second, added to the effects of the rain on adhesion and engine performance, magnified the lack of information that he had regarding the behavior of the new car. Magnesium, which was chosen to

Figure 1.9 **Position of the driver in the Honda RA301 (top) and the RA302**

Table 1.1 Basic characteristics of the Honda RA302 and the Ferrari 312 F1

Model	HONDA RA302	FERRARI 312 F1
Engine	2987 cc (120° V8) air-cooled	2990 cc (60° V12) water-cooled
Compression ratio	11.5	11.8
Ignition	Electronic transistor	Single plug with two distributors and two coils
Maximum power	335 kW at 9500 rpm	306 kW at 11,000 rpm
Mass	Under 500 kg	523 kg (with water and oil)
Tank	200 liters	182 liters
Structure	Light alloy with a magnesium-skinned monocoque	Double aluminum panels joined to a tubular steel structure
Wheel base Front/rear track Length Height	2360 mm 1500/1415 mm 3780 mm 816 mm	2400 mm 1550/1557 mm 4050 mm 875 mm
Suspension	Double wishbone	
Transmission	5+1 speed	
Brakes	Disk	
Distribution	4 valves per cylinder	

reduce the mass of the one-seater, is highly flammable in addition to being light. When Joseph Schlesser lost control of the car, roughly 200 liters of fuel, the high engine temperature and the magnesium caused a severe fire that made it impossible to save his life. Table 1.1 compares the main characteristics of Honda's solution with the Ferrari solution that won that Grand Prix.

Although the other Honda competing in the Grand Prix, the RA301, won second place, the fatal accident suffered by the RA302 caused the Honda team to withdraw from Formula 1 racing from 1969 to 2005.[12] An analysis of this situation in terms of the design process described earlier would begin by examining the motivation that prompted Honda to use the RA302 in that Grand Prix. It would then examine the role of the evaluators who failed to prevent the adopted solution from racing on the circuit. Was the RA302 really the right response to that motivation?

Is it possible that the company's motivation did not match the initial motivation that generated the solution? Is it possible that the solution was not as good as it initially appeared? Is it possible that the unforeseen circumstances affected the solution in such a way that it did not prove reliable? Why did an innovative car such as the Honda RA302 not win the Grand Prix?

This example illustrates the main problem with formulating a design that must satisfy a motivation. How can we know if the solution is good in the initial stages of design, before most of the resources have been consumed? How can we know if one solution, chosen in response to a particular motivation, is better than another? Furthermore, if the actual response can only be known once the design is completed and tests have been conducted under different environmental and operational conditions, often largely uncontrolled or unexpected, how and when should we proceed with the design so that the product will be reliable under all such conditions? Products designed to maintain their performance regardless of possible variations in operational and design parameters are known as robust products, and the design process capable of achieving such products is called robust design. This gives rise to a second[13] need associated with the motivations: *robustness*. When a design is not robust, i.e. when it no longer responds to the motivation for which it was created due to variations in certain environmental and operating conditions, this results in less satisfaction or higher resource consumption.

If we adopt the definition of 'value' shown in Eq. 1.1 to define a 'good solution,' we could rephrase the question as follows: how can we know whether maximum satisfaction and minimum resource consumption will be achieved? This question, in turn, leads to another: in order to achieve minimum resource consumption, how can we know whether the solution will be satisfactory enough before consuming the resources required for a detailed design?

Due to its high degree of abstraction, this is one of the last problems that man has addressed with a general approach outside of a specific discipline. The most usual response in these situations is to think that objectivity cannot be established when faced with a contingent situation, highly dependent on uncontrollable external factors, and that the solution can therefore not be evaluated in advance. The new response is to accept that there are indeed objective patterns that, if not met during the design phase, can be used to catalog the solution as 'bad' prior to completion. Consequently, one of the goals of the design process should be to evaluate the degree of compliance with these objective patterns for each solution. The goal of Design Science is to seek these objective patterns.

1.4 Background

Since the beginning of human activity, design has been a source of wealth. For this reason, mankind has devoted resources to properly exploiting it: abstracting the elements common to the good practices of all those experts was always part of the creation of knowledge in every age.[14] However, changes in mentality are as important as they are slow. Just as physical laws were sought and found by those who believed in their existence, the laws governing the proper execution of a design will be found by those who assume that they exist.

Although the contributions to this process are innumerable (and the problem has yet to be closed), there are two approaches that provide an appropriate frame of reference for the purpose of this book, due to the elegance of their formulations and the power of their results. We owe the first to Genichi Taguchi[15] (1924-), and the second to Nam Pyo Suh[16] (1936-). Both solutions are the product of a common motivation, but at different times and places.

Genichi Taguchi was one of the first to accept the existence of a universal decision criterion, and consequently to seek it. He approached it in terms of a metric that skillfully related the customer's needs, resources, society and the laws of engineering. Thus, he used mathematical sciences, specifically algebra, calculus and statistics, to find a quadratic form that would serve as a tool for measuring the distance from any design to the optimum design. The result produced is the economic loss that society suffers for having placed a production on the market. Taguchi's metric clearly established the relationship between quality and costs. It therefore favored the improvement of production processes through cost savings, noise immunity and improvements in customer satisfaction. This method seeks to improve quality in product design and manufacturing processes, and achieves its greatest successes when combined with statistical experimental design techniques. The results are robust, flexible designs that produce the greatest benefits in large serial productions. Among the competitive advantages attributed to this approach is the reduction of the times associated with research and development during design. Some success case studies have been compiled by Taguchi and Wu (1989).

Although Taguchi had formally solved the problem, as we will see, he had only found a first-order solution with a mathematical formulation suitable for improving completed processes and products, but difficult to execute in the case of new designs. Consequently, he did not have a universally practicable rule, although it was universally valid. This is because the main effects that all engineers must consider when adopting compromise solutions among various objectives, i.e. the main effects that an engineer must assess, were covered in his work. In this way, he could distinguish between two designs. If the result of the metric differed sufficiently between the two, he could discard one without much possibility of error. His metric was used to optimize small aspects of products, fundamentally production processes. In a way, he was responsible for his country's strong technological expansion. Nonetheless, although he had solved the problem, his solution was not directly applicable for addressing a complete design process from the beginning, but rather for making small improvements in a product already on the market. The main problem resides in the algebraic formulation, which requires knowing the input numeric arguments, which are only known if the detailed design has already been completed. To find the best design this way, an infinite number of detailed designs must be made in order to finally select the one whose transfer function and parameters produce the shortest distance to the optimum design. Obviously, strict application of this method during the design phase is impracticable because it would require extremely high resource consumption (infinite, considering the infinite solutions to a particular motivation). Taguchi was aware of this problem, and devoted most of his research to finding a statistical tool that would allow him to make practical use of his metric. This tool was the design of experiments and the associated orthogonal matrices. His work was so extensive, and the number of contributions he made to this branch of mathematics was so high, that experimental design and the orthogonal designs he developed were taken as the key to his success. His fundamental findings were misinterpreted. Even today, in certain areas of engineering, it is not recognized that the added value that Taguchi imprinted on his business activities ultimately resided in the metric he created, and the new definitions of the terms he used.

Having accepted that Taguchi had formally solved the problem (in other words, had found a theory that made it possible to establish a decision tool), Nam P. Suh addressed the same problem around 1980 with a completely different approach. Given that due to the synthesis process itself, there is almost no useful information at all during the initial phases of the design process, Nam P. Suh understood that the approach could certainly not be algebraic. While searching for ways to achieve objectivity without making use of the well-established numerical sciences, he turned to the axiomatic formulation of Design Science. His approach was as follows: let us accept that there is a universal science that says what is right and what is wrong during the design process, and not only at the end. There must therefore be a set of rules that can be derived from some basic principles using formal logic. In this way, he studied what basic principles should be imposed in order to construct the science of design. These basic principles are axioms and, as such, cannot be proven. A peculiarity of the axiomatic approach is that the results can be used in logical argumentations, i.e. in contexts that are more qualitative than quantitative. This characteristic makes Axiomatic Design a fundamental tool in the first stages of design, when the information used is less precise, and there may even be a complete absence of numbers. The fundamental change resides in the fact that the new tool is no longer used to evaluate the final product, but to evaluate the decisions made during the creative process that will lead to it. Today, two axioms are maintained. These two axioms can be quickly stated as follows:[17] 1) Maximum independence, and 2) Minimum information. For example, a possible logical inference from these axioms would be that if all else is equal, the design using the least information content is the best. Once the axioms are fixed, the theorems following from them are rigorously correct. For example, use fewer tolerances, fewer parts, standard parts, symmetries, etc.

Among the many successful results of Axiomatic Design, the development of microcellular plastic, for example, is particularly worthy of mention Youn, J.R and Suh, N.P (1985).[18] In these plastics, the amount of material is reduced by creating pores inside the plastic. Because less plastic is consumed, the cost is also lower. However, contrary to what one might expect, the mechanical performance of the part obtained can be even better than the original.[19]

1.5 The scope of design theories

Because the formulation of each design problem is highly dependent on the context in which it is formulated, it could be argued that these design tools, precisely because they are universal, lack value.[20] In other words, every design problem requires its own design process and its own design rules. Let us consider a simple problem, for example, a chair. Physical laws dictate that at least three legs are needed, but do not say a word about the maximum number of legs. Cost reduction recommends using less material, and therefore calls for a few slender legs. The stability of the chair requires a larger number of legs, and their minimum thickness cannot be too low. Eventually, the first person who decides to put four legs on a chair will seem like an expert, having reduced the number of cases to be considered in the detailed design, and also contributed to the success of the product. But what was the real reflection leading to that decision, rather than considering three or five legs? It was undoubtedly a skillful, subjective consideration of stability, costs, ease of use, ease of manufacture, and a long list of other factors that this person compiled, quite possibly without making any calculations.[21]

The difficulty of the decisions associated with much more complex products in the field of engineering, such as a rocket, a satellite, or an aircraft, makes it necessary to use calculations and considerations simultaneously. In engineering, the number of design parameters to be determined is usually several orders of magnitude higher than the number of physical laws available to determine them. If a design is determined by selecting 200 parameters, and only 20 physical equations are available, how can the remaining 180 parameters be set? By an expert, obviously, who would select the values for those 180 parameters and let the people responsible for the calculations use physical laws to determine the other 20 parameters in order to achieve what the customer needs. But what happens if the 180 parameters are not chosen well? This would be a waste of the resources consumed in the detailed design to ensure the satisfaction of the physical laws. This makes the presence of experts even more necessary. However, there is nothing to ensure that a better selection of the 180 parameters does not exist. Without the help of universal decision tools, this process would undoubtedly depend on the expert, and is therefore subjective: it depends on the training, experience, will, initiative, creativity, etc., of the group of experts. If our group of experts does not have a wide range of prior experiences, the probability of finding a better product is low. (Imagine a situation of complete innovation, where no prior solutions can be used for reference, and the initial design of a revolutionary concept is being discussed, which will eventually lead to a detailed design.)

To fill in the missing information until the expert arrives, we usually turn to our customers. Over time, customers have developed new needs, or rather, they have become more demanding. Thanks to the abstractions based on the multiple decisions made by experts over decades, they have learned what to ask for to ensure additional benefits for the product. These needs are new constraints that have been added to the formulation of the problem: stability, contamination, mass, volume, reliability, robustness, safety, consumption, etc. This whole list of needs imposed by the customer must be completed with the internal motivation of the designer: to achieve the best design. In other words, to satisfy the customer's needs with the best product possible in order to attract new customers.

Design theory is comprised of that which is common to the infinite aggregate of experts working for an infinite set of customers. The scope of a theory of design must be sufficient to enable correct decision-making on at least the following aspects: 1) Finding the best definitions of the needs, either in terms of nominal values to be satisfied or merit functions to be optimized, or in terms of a particular formulation that must be made explicit. 2) Establishing a solution that makes it possible to obtain the nominal or optimum values of the merit functions by varying and controlling a series of design and operational parameters, i.e. establishing the best transfer functions of those predicted by the system engineering for each solution. 3) Establishing which design parameters must be adjusted or controlled, and which must be avoided and fixed by tolerances, i.e. cataloging design parameters as better and worse.

The way the problem has been approached, the basic need that a design theory must satisfy, is to build the optimum transfer function and select the optimum values for all parameters so that the best solution is found objectively. In order for this to occur, it must produce new equations. These new equations do not come from engineering, physics, or the discipline in question, although they are related to them. Rather, they are equations originating from the design theory itself. They are therefore design equations available only to those designers working with that design philosophy. The new equations generated by the design methodology are the closing equations that complete the problem, so that there are as many equations as there are unknown quantities. Obviously, the more universal and objective the design theory used is, the better the final set of design equations will be. In this sense, the design process capable of generating the set of equations whose only solution is the best will be the optimum design process. The goal is to find the theory capable of supporting such a process, referred to in this book as an *advanced design process*.

Obviously, the engineering equations will be active at the same time as the new equations generated. Design science will never be able to create equations from physics. The idea that Design Science is sufficient for producing good designs is therefore completely mistaken. On the contrary, a multidisciplinary group of excellent engineers is required to make adequate use of the new tools arising from Design Science.

1.6 The definition of design

Although it may seem obvious, the design is the result of the design process. For example, it is the set of real products manufactured and established by the design process, but it also includes the very set of abstractions and decisions involved in the process. The broadest definition that includes the peculiarities[22] described above is:

Definition. To design is to formulate and execute a plan to satisfy a need.

This definition includes the verb 'formulate,' the action through which an order, proposition, need or, in this case, a plan is reduced to clear, precise terms. By plan, we understand the document specifying the details for creating a work. A work is any intellectual product in the sciences, letters or arts. The action 'satisfy' refers to the objective, the purpose for which the plan is formulated, and specifies that the objective is to meet the conditions expressed in a problem, thereby providing an acceptable solution to it. The word 'need' refers to those conditions that make it possible to define the design problem, and clarifies that they are presented as an irresistible impulse, a motivation, causing the actions to move infallibly in a certain direction. The verb 'execute' allows no possibility for the formulation to lose contact with reality, and forces us to take into account the laws that condition the behavior of this reality.

This is the definition adopted because it establishes design as a process in which the result is oriented towards an objective and subject to restrictions. For example, the condition of reaching the formulation in clear, precise terms is a restriction, while the conditions imposed by the satisfaction of a list of needs are an objective. We will now compare this definition with others that differ slightly.

Example 1: It is the formulation of a plan to satisfy a need. This definition does not require execution, and can lead to untested formulations. The formulation is made to satisfy a need, but if it is not executed, the degree of satisfaction achieved is not known.

Example 2: It is the process of applying different techniques and scientific principles in order to determine a device, process, or system in sufficient detail so that it can be created. There are two serious problems with this definition. First, it does not explicitly specify the existence of a customer whose needs prompt and direct the process. And second, because the needs are not specified, it also does not indicate how designs should be evaluated in order to distinguish the good from the bad.

Example 3: Design is the process in which scientific principles and technical methods (knowledge of mathematics, physics or chemistry, drawing and calculation tools, common or specialized language, etc.) are used to carry out a plan that will result in the satisfaction of a particular need or demand. This example corrects the problem with the previous definition, but makes the mistake of excluding those acts that are purely creative. Such acts, by definition, do not fit within the scientific framework. By not regarding creativity as science,[23] this definition eliminates those plans provided by experts whose experience has led them to develop a sound, prodigious intuition for solving the problems that concern them. In any case, this definition does not include the creativity, discovery, thought and intuition that are part of the creative process, which, in turn, is an inevitable part of the design process.

Example 4: To design is to define.[24] This definition is based on a change of terms. The problem of explaining what it means to 'design' becomes the problem of explaining what it means to 'define.' Although nothing is said about the objective of designing or defining, or the purpose of what is defined, it could be understood that to design is to define the design problem and the solution, both in terms of needs because they give a purpose to what is defined. By following this path, we would end up with a definition of design very similar to the one adopted: design is the formulation and execution of a plan to satisfy a need. However, as we will see in Chapter 2, this definition is very interesting if we assume that 'define' means 'to reduce uncertainty.'

1.7 The characteristics of design

We can see that all of the characteristics listed below are compatible with the definition given above:

1. All design problems are always subject to certain conditions in order to be solved.[25]
2. The degree to which the different conditions are satisfied is an indicator that enables us to decide which response to the design problem is better.
3. A design problem is a decision-making problem.
4. A design problem requires the creative handling of a large amount of information in order to reach a satisfactory solution.
5. A design problem is not a hypothetical problem, but a real one. Its specific purpose is to obtain a final result, which is achieved by performing a particular action or by creating something that has a physical reality. This is always the case because the needs are met on the physical plane.
6. Design problems do not have a single response.

All of these characteristics give design theories a high degree of abstraction. However, the last one in particular is what makes them contribute value. The existence of multiple responses comes fundamentally from the creative process. Creativity, necessary for proposing and generating solutions, means that the design process has a subjective component. Of the possible solutions found, one of them may be optimum, but nothing is known about whether the best solution is among the set of solutions proposed. In general, an optimized solution is not necessarily the optimum solution, just as an optimum solution is not necessarily the best solution. This book postulates that the best solution always exists. In the context of this book, we will refer to the best solution as the *ideal design*. The ideal design is achieved by systematically applying advanced design theories.[26]

1.8 Design problem

A design problem is born when the customer's motivation is stated in terms of needs. However, the description of the needs often contains part of the information necessary to create the solution that the customer considers to be most feasible.[27] Thus, we often find statements such as the following: 'What we need is a gizmo we can stick into this thing to close a hole only while this machine is operating within the indicated time.' We have to ask ourselves whether what is really needed is a gizmo, or simply for the hole to be closed during the specified time periods. It is very important for design problems to be formulated with the lowest number of conditioning factors possible. However, it is not easy to find the hierarchical structure of needs in order to isolate the dominant one.[28] Normally, the hierarchical structure does not exist a priori, or has been defined based on circumstantial criteria that should be carefully examined. In general, we should accept that the hierarchy of needs is not well structured. For example, the solution to the problem 'I need a car' is completely different from the solution to the problem 'I need to get around.' In the first case, the first level of the hierarchy of needs already assumes that a car is the solution. In the second case, however, the solution may or may not appear as a reflection of the need. We might then ask ourselves whether the true motivation has not been distorted by the presence of an existing solution. In such a situation, we have no choice but to structure each problem to check that the motivation was the appropriate one, or to find the true motivation: we must provide the *available information* that will allow us to correctly reinterpret the motivation.

To structure the hierarchy of needs, we must perform an *analysis* (break down, separate, break up into parts and components). However, in order for us to analyze the problem, it must be completely defined, which is why it is often necessary to adopt a possible solution. To define this solution, we must first perform a *synthesis* (organization, integration). Even without defining a solution, in order to understand the same motivation (process of analysis), it may be necessary to generate (process of synthesis) a list of needs. (Notice that to establish one list of needs rather than another is to adopt a solution to the problem of defining the motivation.)

Thus, several complexities of the design process arise. How can we analyze something that has not been created yet? How can we synthesize based on parts that have not been created? The designer must take a blank sheet of paper and fill it with ideas. Obviously, the ideas proposed will be biased by the design problem. Two different statements of the same motivation will lead to different sets of ideas. This is why determining the tools needed to establish a correct approach to the problem is also part of design science.

Naturally, this process takes place continuously. In particular, when product specifications are specified through a contract, it is essential to reformulate the contract until the requirements established as necessary are clearly specified in order to reduce future resource consumption. Some kind of assessment criterion is needed for this purpose, just like when the analyzed actions are compared to the requested needs during the design process.

1.9 Activities in the design process

Once an objective has been identified and established, any design must continue with a creative process, i.e. a decision that places a possible solution on the design table. At this point, an analysis[29] can begin in order to generate solutions better than the initial one. This process transforms the tentative initial solution into the final solution adopted through successive iterations. Table 1.2 provides a list of 10 activities representing a possible linear description of the process.

There are many other descriptions of the design process, with a higher or lower number of activities and phases.[30] Nonetheless, for the introductory purpose of this chapter, the description comprised of two phases with five activities each will be sufficient. Each of the activities

Table 1.2 Typical activities in a design process

Phases	Activities		
Synthesis	1.	Identify the need	MANAGE THE INFORMATION
	2.	Collect information	
	3.	Establish the objective	
	4.	Specify tasks	
	5.	Create tentative solutions	
Analysis	6.	Assess the tentative solutions	
	7.	Evaluate the tentative solutions	
	8.	Detail the selected solution	
	9.	Create prototypes	
	10.	Produce	

shown sequentially in the table consists, in turn, of another set of activities. Thus, the definition of each phase could reach a high level of complexity. While not intended to be exhaustive, the following is a description of some of the steps comprising the main activity of each phase.

1.9.1 Activity 1: identify the need

This phase normally begins with an unstructured approach to the design problem, in which the formulation usually comes from the customer. In order to structure the problem to the highest degree possible, the different needs involved are identified. The dependencies and constraints between them are determined. This information is used to create a hierarchical list of needs. Based on the hierarchy of needs, new groups of customers can be identified, both current and potential.

1.9.2 Activity 2: collect information

The maximum amount of information must be obtained regarding the needs involved. This is a thorough investigation aimed at collecting information on physical, chemical and other relevant aspects of the problem, existing solutions for satisfying each need, and similar or related products on the market. The available solutions are assessed. If a satisfactory solution already happens to be on the market, it will probably be more economical to buy it than to create another one. The state of the art for similar technologies and products is established. Available patents and technical and specialized publications are studied.

1.9.3 Activity 3: establish the objective

Express the problem again with a more coherent approach. This new specification of the problem must be a functional visualization with three characteristics: it must be clear, concise and general. All references to a particular solution must be eliminated. This way, it will not be nuanced by terms that predict a solution, which could unnecessarily limit the designer's creativity.[31] The goal must be stated correctly in terms of the list of needs and the state of the art.

1.9.4 Activity 4: specify tasks

Find the operational specifications. Their mission is to restrict and define the problem, serve as a contractual definition of what must be achieved, and ensure that the finished design can be evaluated in terms of compliance with these specifications.[32] The tasks required for the correct execution of the design must be established, as well as the time frames, the people responsible, and the resources that will be allocated.

1.9.5 Activity 5: create tentative solutions

Devise. Invent. During this step, possible solutions are considered, and different techniques can be used. The creative process involves generating ideas, avoiding frustration, promoting the incubation of new solutions, fostering 'Eureka!' type situations, iterating from Activity 1 (identifying the need). To promote the generation of ideas, it is advisable to defer judgment or, in other words, temporarily suspend one's critical spirit.[33] At this point, a synthesis has been made, resulting in a series of possible systems on which an analysis can be performed.

1.9.6 Activity 6: assess the tentative solutions

Listen to experts, engineers and customers. Manage teams with experts from various disciplines.[34] Produce or purchase assessment tools. Promote the use of objective assessments. Establish assessment criteria. Assess. Make hierarchical lists of the assessments of the solutions.

1.9.7 Activity 7: evaluate the tentative solutions

Evaluate and judge in order to select a possible solution. Of all the systems deemed viable after the prior analysis, choose one. Establish judgment categories. Establish decision matrices with the possible designs in the rows, and the judgment categories in the columns. Weigh each category.[35] Decide whether the synthesis phase should be iterated, repeating the process from Activity 1 if necessary. Assess the impact of bifurcations on the design flow.

1.9.8 Activity 8: detail the selected solution

Perform the system design. Perform the detailed design. Design the parameters and tolerances. As a result, a complete set of drawings and assembly and detail diagrams will be obtained for each and every one of the parts used in the design. Each detail drawing must show all dimensions, tolerances, materials, treatments, etc.

1.9.9 Activity 9: create prototypes

In addition to mathematical and simulation models (also very useful for assisting with design during phase 8), it is advisable to produce a real prototype. The prototype might be built to scale or full-sized with simplifications, depending on the budget. However, the ideal scenario is ultimately to produce the final products, just as they will be used, and to establish a series of significant tests to run on them.

1.9.10 Activity 10: produce

Design the manufacturing, assembly and other processes. Once all of the above activities have been completed, problems may arise with product production. To avoid this, we can turn to Concurrent Engineering,[36] where product and production engineering meet to provide an optimum solution for the product and the process at the same time.

1.10 Information management

Communicating the final design to other people is the final, vital step in the design process.[37] Not only is the final information relevant, but also the information generated during all of the intermediate steps. This information enables us to make decisions leading to continuous product improvements. Implicit in information management is a decision-making system, which is described in Activities 6 and 7 of the design process itself. Such information management produces the necessary feedback between the different activities in the two phases, which are the so-called iterations of the design process. Not only are the synthesis and analysis steps related to each other through an iterative process, but it is normal for several of these processes to take place in parallel, such as Activities 8 and 9. In any case, without including feedback or parallel steps or intending to be exhaustive, the following figure illustrates the level and quantity of information that must be managed.

Thus, information management is a transverse activity because it relates the inputs and outputs of each and every one of the activities

Figure 1.10 Example of a possible deployment of the information-activity-information pattern in a design process

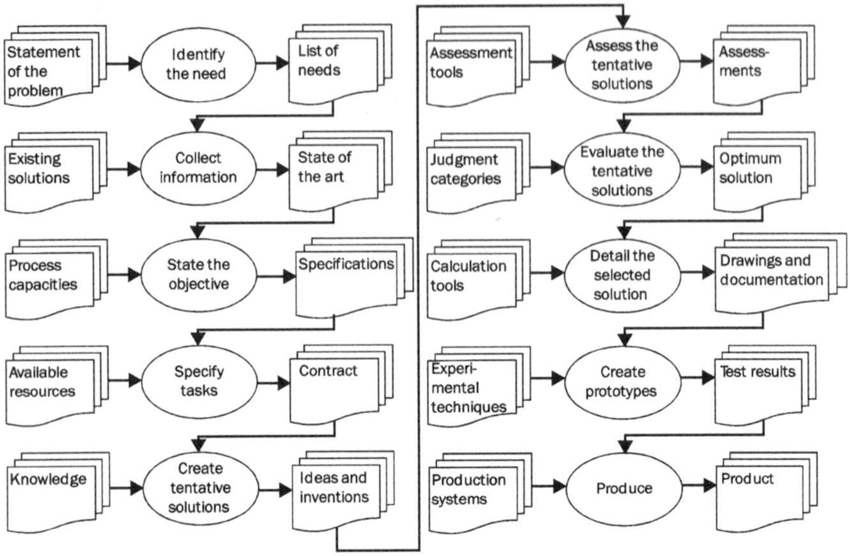

carried out during the design process. Correct information management is obviously a necessary requirement of any good design process.

In a general case, all of the synthesis and analysis activities are clearly interrelated through the *iteration* that takes information from a later activity in Table 1.2 to a previous activity. In a general case, it may be necessary to skip from one point to another of the process as many times as necessary. Theoretically, such an iteration, or information feedback, could continue infinitely for a given design problem, continuously creating small improvements. Inevitably, the incremental gains or cost reductions achieved will tend towards zero over time. This allows us to define different stages depending on the information content they include. For example, a possible division could be:

1. **Conceptual or preliminary design:** The objective is to choose one of all the possible solutions. Most of the main characteristics of a design are determined during this stage, which usually represents a small fraction of the total time occupied by the design process. This stage encompasses up to step 7 (selection). The information output is the optimum solution or frozen configuration.
2. **Detailed design:** With most of the important decisions clearly established in the preliminary design, we get what could be called a

'frozen design.' This is the time to create engineering (mathematical) models that enable us to analyze the elements of the system and the system itself. Because these models are complex and expensive to create, it would be desirable if the frozen design did not require modifications following the detailed analysis performed to fine-tune the parameters. The information output is the set of drawings, manuals and instructions that enable us to perform the tasks leading to the placement of an operational product on the market.

Obviously, real situations are quite a bit more complex. This makes it necessary to establish multiple phases, as we will see below.

1.11 The design process as a product

The sequentiation shown in Table 1.2 and explained in points 9 and 10 is no more than a tentative solution to the design process. The following questions arise: Is this the optimum design process? Is there a single optimum design process, or does the optimum process depend on the type of product designed? Is it advisable for the creators and the analysts to be independent?

There are multiple answers to these questions. For example, the European Space Agency (ESA) and other international agencies recommend dividing projects into different phases. The definition phase, subdivided into phases A and B1, is followed by the implementation phase, comprised of phases B2, C and D. Then comes the operational phase, phase E, and finally the disposal phase, phase F. Phase B2, the first implementation phase, is the preliminary definition phase. Its results are reviewed and assessed in the PDR (Preliminary Design Review). This review is performed on all elements comprising the different levels of the project, covering the space vehicle, the payload, the rocket and the ground segment. Following the PDR of each element, a PDR is conducted at the mission level (Mission PDR). The detailed design of all elements is performed during phase C. During this phase, the qualification tests necessary to support the design are also conducted, and the scheduling of the next activities is improved. Phase C culminates in the CDR (Critical Design Review). This review authorizes the manufacture, assembly, integration and testing of the device that will finally be launched.

In 1998, ESA also started using the CDF (Concurrent Design Facility) tool at ESTEC, in an attempt to apply concurrent engineering[38] to the assessment of space missions during conceptual design. Three years later

(Bandecchi, 2001), the main objectives of the tool seek to increase its capacity to: 1) generate a complete environment for the conceptual design of new space missions,[39] 2) fine-tune the application of the principles of concurrent engineering, 3) more efficiently organize the design tools and human resources dedicated to mission analysis, and 4) capture and retain knowledge within the corporation for later use.

When knowledge is later used to redirect the process by modifying decisions adopted during the previous steps, this is feedback. It is remarkable that this same tool presents the feedback by repeatedly executing the same phases each time a cost analysis is concluded. For example, a spiral model can be used to illustrate the iterative process. In the case of a space mission, it is outlined as shown in Figure 1.11.

The main ingredient that makes *concurrent design* viable is correct communication between the multiple disciplines. To ensure this, directed sessions are held. Each plenary meeting brings together representatives from all space engineering disciplines and customers, under the coordination of a team leader. Among the disciplines invited are: systems, instrumentation, mission analysis, propulsion, attitude and orbit control,

Figure 1.11 **Spiral diagram illustrating the repetition of phases in each iteration of a possible design process in the aerospace industry**

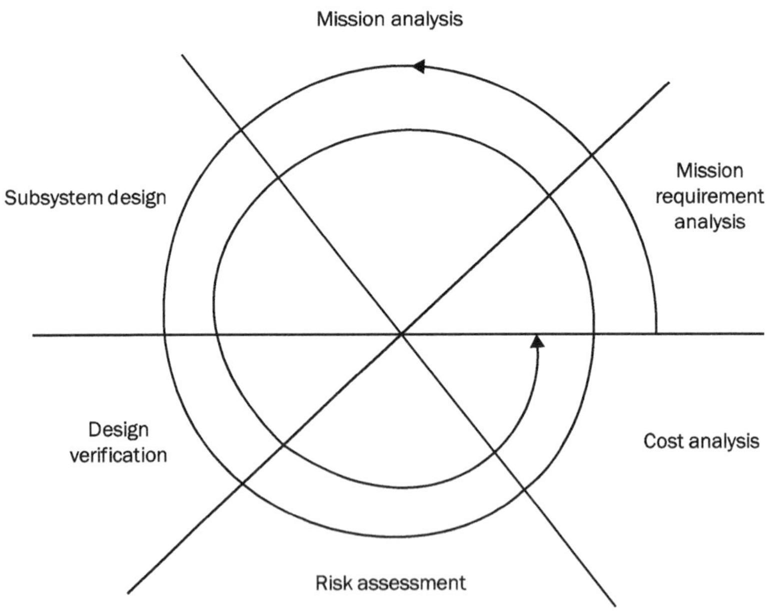

Source: Bandecchi, 2001, Courtesy of ESA.

structure, configuration, mechanisms, thermal control, power, data and command management, communication, ground system, operations, simulation, programming, risk assessment and cost analysis. It is remarkable that participation in these meetings is maintained from the initial phase (requirement analysis) to the final phase (cost analysis), and particularly notable that at least one senior and one junior representative from each discipline is present. The tools must enable model-based design, online design, cooperation and interaction. Under these conditions, a study prior to phase A can mean a preparatory phase lasting two weeks to two months, a concurrent design phase typically lasting three to six weeks (6 to 10 sessions lasting four hours each, taking place every two weeks), and a final documentation phase lasting one to two months.

In addition, all of these activities take place in a context in which ESA must work with the industry in charge of creating the subsystems. For this reason, a typical design context at ESA might be as shown in Figure 1.12.

1.12 The importance of the design process

All of the above points reflect some aspect of the design process. All of these aspects are undoubtedly different, yet related. It is a dynamic process involving time and a multitude of objects that evolve, and whose evolution, in turn, conditions future evolutions. Without a doubt, this is complexity. Mankind's activities aimed at creating value and satisfying needs are composed of a multitude of tasks that handle a multitude of objects in a very complex context. This complexity is also subject to variations, noise and unforeseen circumstances, whose impact must be controlled.

It is easy to understand that a product's success lies first in the success of the design process. The success of the design process is a necessary (and sufficient?) condition for the success of the product. If the product is not a success, this is because it does not sufficiently satisfy the needs, and was therefore badly conceived, or badly designed.[40] This proves the necessary condition. To prove the sufficient condition, let's imagine that the design process was successful, a plan was formulated, and it was carried out in a way that satisfies the needs. Consequently, the product is successful in that instant, but it might no longer satisfy the conditions an

Figure 1.12 The relationship between industry and the European Space Agency in a typical design process

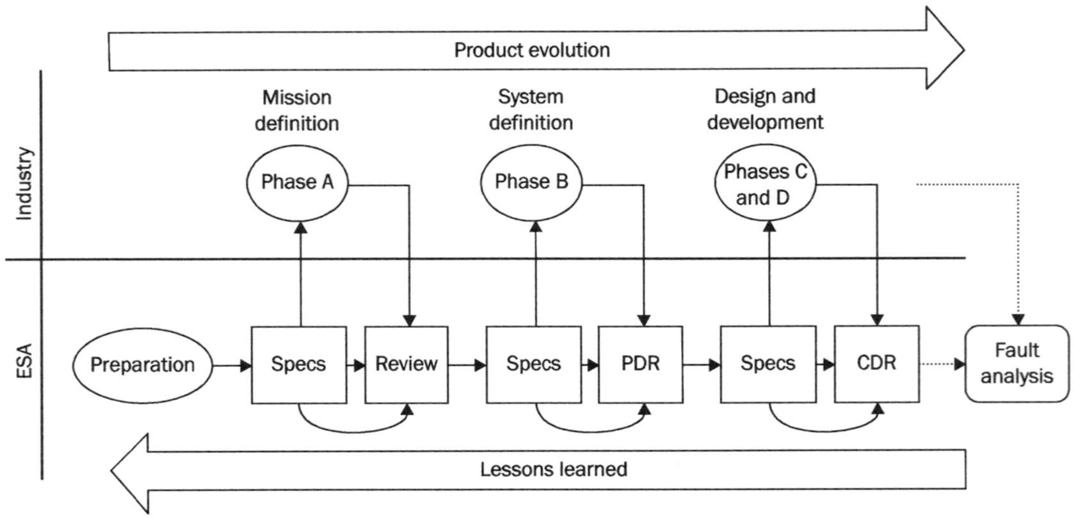

Source: Bandecchi, 2001, Courtesy of ESA.

instant later, for example, because a better product comes along. The concept is therefore dynamic, and the time that a design remains on the market, satisfying the needs, is a measure of how good the design process was. This would indicate that maximum product lifetime should be included as a need. We could apply the same reasoning to costs, etc. When all of these needs are included, it becomes clear that a poor design process[41] implies a poor product. These initial reflections indicate the path to follow, and its difficulty. A good product starts with a good design process, but how do we design a good design process?

1.13 The importance of Design Science

It is time to acknowledge the importance of Design Science, and to answer the question raised in the previous section on the origins of Design Science.

1. Design Science can be understood as the body of knowledge obtained through observation and reasoning, systematically structured, from which general principles and laws are deduced.
2. Design Science can be understood as skills, mastery, or the body of knowledge in any area.
3. Design Science can be understood as the body of knowledge relating to the exact, physical–chemical and natural sciences.

These three meanings, reformulated based on the most common definitions of the word 'science,' summarize very clearly the complexity referred to throughout this chapter. On the one hand, they encompass all prior knowledge of the art of design, and structure it in order to extract that (minimum) set of general principles and laws leading to the objective of the science: making correct designs. On the other hand, they cover the artistic, creative aspect necessary in any process of synthesis, through skill, mastery and multidisciplinary knowledge (knowledge in any area). Finally, contact with the real world is reflected in the necessary knowledge of the exact sciences, i.e. mathematics, to include the principles and laws governing nature and physical–chemical processes.

Thus, Design Science in general, and advanced engineering design in particular, necessarily rests on these three pillars. Of the three, only the first is completely transverse and objective.[42] It generates added value because its transverse nature allows it to be applied to all phases of the design process, and establishes a language common to all disciplines. In other words, it establishes the common framework in which the different branches of engineering must coexist so that the final result will be sustainable, optimum, excellent, or of maximum quality, depending on the forum where it is applied.

1.14 Notes

1. The concept of available information is discussed in Section 2.12.
2. For this reason, it is often said that the proper approach to a design problem is fundamental. During these stages, it is usually advisable to use tools such as QFD (Quality Function Deployment). For a description of QFD, see Shillito (1994).
3. The design problem formulation provided by Axiomatic Design (Suh, 1990) uses a list of functional requirements and another list of constraints. This formulation requires choosing a minimum number of independent functional requirements, and therefore adds value as opposed to a mere description through a list of needs. Thus, the separation between requirements and needs is an issue of particular relevance, which must be handled carefully while teaching Axiomatic Design (Brown, 2005).
4. Efficiency acquires a universal formulation when discussing sustainability. In private contexts it is formulated, for example, in terms of business costs or thermodynamic performance, depending on the working environment.
5. All disciplines are characterized by a set of objects that obey certain rules. In the case of art, those works that produce openness to introspection constitute a branch of modern art. In engineering, the set of devices capable of lifting off the ground while transporting a payload constitute aerospace engineering. The laws of color and texture in painting, and the physical laws in engineering, are rules of these disciplines. These rules are introduced into the process by the creators and producers, and are restrictions that limit the set of possible solutions.
6. Some of them will be responsible for cataloging works of art, e.g. by specifying those considered masterpieces, and others will be in charge of cataloging engineering works, e.g. by establishing the qualities of industrial products or their ease of use.
7. Physical laws impose objective constraints; however, these have already been considered by the creators and producers while creating and materializing a possible response, and are almost always transparent to the evaluators.
8. The decision to use value, defined as the ratio of satisfaction generated to resources consumed (definition adapted from Hundal, 1997), is itself an arbitrary or subjective decision because another choice could catalog another initially discarded solution as better. It is therefore essential to objectify the indicator capable of establishing the scale that differentiates good designs from bad. That is one of the fundamental objectives of this book, which will be developed in later chapters. For a more detailed discussion of the concept of value, see Dean (1993b, 1995). Also, because the expression satisfaction/

resources is measuring the value of the solution in some way, it seems logical that the basic units of measure should be monetary units. However, this is not the case due to the great difficulty involved in defining intangible factors such as the degree of customer satisfaction in economic terms. One possible solution to this problem is G. Taguchi's quality loss function, which will be discussed in Chapter 4.

9. Intuitively, it seems logical to assume that the appropriate design is the one that maximizes the value. Nevertheless, other possible criteria will be established throughout this book: 1) the appropriate design is the one that verifies the axioms, or 2) the appropriate design is the one that minimizes the loss of quality.
10. For more on this issue, see Glegg (1960), Harrisberger (1966), Hill (1970), Ostrofsky (1977), Simon (1969), Suh (1978), Suh (1984), Yoshikawa (1985), Hubka and Eder (1987) and Suh (1990).
11. John Surtees participated in the 1968 French Grand Prix driving the Honda RA301, an earlier version of the RA302. He finished 2nd after Jacky Ickx, who was driving the Ferrari 312F1.
12. During this period, Honda did not lose contact with racing, as it continued to supply engines and chassis to other racing teams.
13. The first is efficiency, described in Section 1.1.
14. Obviously, this eventually resulted in a stable doctrine that is taught at universities, which transmits what is considered good for the discipline, and does not transmit what is considered bad. This is proof of the existence of decision criteria with a high degree of objectivity.
15. Throughout this book, the approach based on G. Taguchi's ideas will be referred to as Metric Design.
16. The approach based on N. P. Suh's ideas is called Axiomatic Design.
17. The precise formulation will be discussed in Chapter 3. Axiom 1: Maintain the independence of the functional requirements. Axiom 2: Minimize the information content.
18. These plastics were the response obtained by applying the axiomatic design methodology to the business motivation of reducing the material used in the manufacture of packaging without losing mechanical performance (Suh, 1996). A description of this and other examples of such applications can be found in Suh (2001).
19. Because pore size can be chosen independently from pore number, if the pore size chosen is below the critical size capable of starting a crack, it is difficult for cracks to spread inside the material.
20. This objection is basically refuted with the very definition of science. If the universality and generalization of cases had not been a motivation in itself, none of the sciences would exist today. The Laws of Newton, which describe the mechanical behavior of our environment, would not exist, but rather an extensive collection of recipes based on a multitude of experiences.
21. To solve this problem, we could optimize a merit function, such as stability/mass (maximum distance from the center of mass to each of the straight lines obtained by joining the bases of the different legs, divided by the mass of the chair), or resort to asking users about their degree of satisfaction with each solution. However, these responses are somewhat subjective. The merit

function mentioned is arbitrary; we could also have used stability/number of legs, or asked a different group of users.

22. Akiyama (1991) discusses some of the peculiarities of design that are implicit in the definition proposed. He argues that design is an activity that recognizes objectives and purposes; gives form to objects, determining them and making them universally understandable; evaluates objects; and transforms the customer's demands into a specific product.
23. At the moment, the theory most widely used for discussing creativity is TRIZ (Orloff, 2006).
24. Dean and Unal (1992) maintain that to design is to define, and that functional analysis and quality function deployment (QFD) are tools for specifying a definition. For Jacobson et al. (1992), design in the context of software development is an evolution from an abstract model with few dimensions to a concrete model with more dimensions.
25. These conditions can include the customer's needs, outside constraints, engineering constraints, and constraints from the design theory itself.
26. By definition, the ideal design, or best solution, cannot be improved upon (at least not significantly). For this reason, products designed in this manner will naturally generate a market entry barrier for new competitors because they will lead the market in quality as well as performance and price (for the same production size). Any differentiation strategy will either be unviable, or will generate a completely new product motivation. The only appreciable improvement to the best product lies in reducing production costs (for example, by changing the production size).
27. In the context of Axiomatic Design, design problems must be approached independently of any solution, in what is known as a solution-neutral environment. An environment associated with a solution may only be maintained when improving upon an existing solution, taking into account the solution attributes that the customer considers most important, or to which he is most sensitive. These are known as the customer attributes.
28. Chapter 6 presents one example where hierarchy is established by means of the advanced design procedure.
29. In the NASA design concept, Edwin B. Dean defines functional analysis as the tool that enables us to study the characteristics of a system. Here is a summary of this approach (Dean, 1997): 'Although the products and services exist as physical objects or systems, they are not created from scratch. They are preceded by an idea – a concept – which is the basis of their creation. Functional analysis identifies the nature of products and services by bringing those concepts to light. Functional decomposition is the process of asking "How?" for each higher-level function, thus deriving lower-level functions. Functional composition is the process of asking "Why?" for each lower level of functionality, thus deriving higher-level functionalities. The result is a tree or systematic diagram of functionalities that hang from some higher level. For engineering systems, the top level of higher functionality is the purpose of the system, and the lower levels are the means for achieving the purpose.' This process is adequately described in the Quality Function Deployment (QFD) technique (Dean, 1992).

30. Dean (1993a) provides an interesting approach regarding phases, people and processes in system design.
31. 'A lawn mower is needed' or 'The grass needs to be kept short' can be different formulations of the same problem, which can lead to two very different solutions. (Solution 1: An electric lawn mower. Solution 2: Feed a flock of sheep.)
32. In the final product sold, the difference between the nominal value of these specifications (the customer's need) and the actual actions taken (the purchased reality) produces dissatisfaction. Along with the price, this will define the quality perceived by the customer.
33. The goal is usually to obtain the largest number of potential designs possible, which is why it is common to find activities conducive to the gathering of ideas: brainstorming in groups of 6 to 15 people, the use of analogies (switching to a physical analogy), the use of inversion (replacing the mobile with the static), the use of synonyms (finding synonyms of the verbs that appear in the statement of the goal). One good tool is the TRIZ theory of inventive problem solving (Orloff, 2006).
34. While advisable during every phase, during this phase sufficient knowledge of the disciplines involved is essential: mechanics, electricity, thermodynamics, fluid dynamics, mathematics, computer science, etc.
35. A weighted value obtained from the user or the experts is usually assigned to each category. For this stage, the use of the QFD (quality function deployment) technique is very useful, particularly in the initial phases of product definition.
36. Jagannathan et al. (1991) define concurrent or simultaneous engineering as follows: 'The process of forming and maintaining multifunctional teams who specify the parameters for the product and the process as soon as possible in the design process.' Dean and Unal (1992) described it this way: 'Concurrent engineering means putting together the right people at the right time to identify and solve design problems. Concurrent engineering means designing for assembly, availability, cost, customer satisfaction, maintainability, manageability, operability, performance, quality, risk, safety, times, social acceptance, and all other product attributes.' Concurrent engineering includes the simultaneous design of the product, product evaluation, creation of product prototypes, product testing, product production, product deployment, product operation, product maintenance, product withdrawal and product management. All of these steps are encompassed by the concept of 'genopersistating a product' (Dean, 1993a). Concurrent engineering usually tends to reduce development time and costs when appropriately applied. Section 1.12 describes the concept of concurrent engineering currently adopted by ESA.
37. It is often said that he who sells a new idea also sells himself as an originator of ideas. True failure would be to completely refrain from presenting ideas, i.e. to eliminate Activity 5 from the design process, because any kind of analysis is not possible without a synthesized object.
38. The definition adopted by ESA (Bandecchi et al., 1999) is: 'Concurrent Engineering is a systematic approach to integrated product development that emphasizes the response to customer expectations. It embodies team values

of cooperation, trust and sharing, in such a manner that decision making is by consensus, involving all perspectives in parallel, from the beginning of the product life-cycle.'

39. It had only previously been applied to the internal phase corresponding to assessment studies, level zero of phase A.
40. For example, the needs might not have been analyzed correctly, so that it was not discovered that the product did not actually satisfy any needs. Or, a fundamental need might have been forgotten. Beta and 2000 video recording systems succumbed to VHS because they neglected a fundamental need. Once the need to record and play images and sound had been satisfied, the next need was for abundant visual material to be available in that format, along with an abundant supply of player/recorders. The successful system was the one that managed to become the standard video format in the shortest time.
41. For example, because it incurs excessive design costs which must later be recouped.
42. The second is transverse, but not objective. It is subjective because it depends on a creative process. Without the idea, there is nothing.

2

Information, entropy and its relationship to design

Abstract: The previous chapter discussed the most important characteristics of the design process. This chapter proposes mathematical definitions for modeling some of those characteristics. It defines the concepts of variable, alphabet, spaces of definition, uncertainty, entropy, mutual information, information content, acceptance margins, tolerances, response margins, and transfer function. The main objectives are to explore the relationships between these concepts, explain the importance of the concepts of entropy and information in the design process, and seek upper bounds for them. As a result, this chapter shows the first general conclusions affecting the planning of a design process: statements #1 to #7.

Key words: variable, alphabet, spaces of definition, probability distribution, uncertainty, entropy, mutual information, information content, acceptance margins, tolerances, response margins, transfer function, available information, necessary information, mediation operator, synthesis operator, calculation operator, evaluation operator, analysis operator.

2.1 The design process in terms of probabilities

The design process can be understood as a process that decreases the degree of abstraction of the objects handled until finally reaching the maximum concreteness when the probability of success[1] is within certain acceptable limits. The decisions made during the design process determine the probability of success because they choose one option and not another. It is natural for some options, which could initially appear very attractive, to have a greater probability of being discarded as the design process progresses. These ideas enable us to describe the design process as a temporal evolution of certain probability distributions. To see this, let's begin with a simple situation.

Suppose that a department has been given the task of generating a set of tentative solutions to a particular design problem, and that it belongs to a larger design division. This department finds n possible solutions. These are named using a set of available labels, for example, the labels comprising the elements in set $O_X = \{x_1,x_2,...,x_n\}$. Obviously, all of these solutions have survived the same selection criteria present during the synthesis phase. There is therefore no clear argument in this department for selecting one over another. Under these conditions, all of the solutions are equally probable. Now, suppose there is a second department within the same design division, whose task is to choose a particular solution from the ones provided by the previous department. To do so, this department will have to use new criteria that were not available to the previous department. Let us assume that, having applied these criteria, they manage to rule out all of the solutions except two, which they believe to be equally feasible, for example, x_2 and x_4. If a third department were to assess and evaluate the two surviving solutions with new decision criteria, it would be able to select the best one; let's say x_2. A decision has been made in each of these steps, first restricting the set of feasible solutions from $\{x_1,x_2,...,x_n\}$ to $\{x_2,x_4\}$, and then to $\{x_2\}$. The decision making has increased the certainty of the solution. It could initially be any value from set O_X with the same probability, and finally only the value x_2 with complete certainty. In general, this process can be interpreted as the evolution of a probability distribution.

Definition: We can say that the probability of element x_i in set $O_X = \{x_1,x_2,...,x_n\}$ is $p_i = \Pr|_{O_X}(x_i) \geq 0$, where $\sum_{i=1}^{n} p_i = 1$, if variable X with alphabet O_X takes the value x_i with the probability p_i. Set $P_X = (p_1,...,p_n)$ containing the probabilities of each element is also called the *probability distribution* of variable X.

With this definition, the above design process, which first generated solution set $\{x_1,x_2,...,x_n\}$, then selected $\{x_2,x_4\}$, and finally chose $\{x_2\}$, can be interpreted as the evolution of the probability of the different elements making up the set: the output from the first department will be set O_X with the uniform probability distribution $P_X(t_1) = (1/n,...,1/n)$. As a result of the work of the second department, the above distribution is transformed into the probability distribution $P_X(t_2) = (0,1/2,0,1/2,0,...,0)$. Finally, the resulting probability distribution will be $P_X(t_3) = (0,1,0,...,0)$, where $t_3 > t_2 > t_1$ are the instants of time in which each department finishes its work.

Note that nothing has been said about the nature of the decision-making tool. These can range from the purely random (for example, flipping a coin) to the most sophisticated, based on the execution of each tentative solution down to the last detail in order to check the degree of satisfaction produced. Obviously, both extremes are inefficient. The first consumes resources to produce a solution that may be far from what the customer expects, and the second consumes resources to produce n-1 solutions that will eventually be ruled out. It is the task of engineering and design science to establish the tools to optimize such decision making; in other words, to ensure the largest amount of useful information[2] for correct decision making with the lowest resource consumption.

2.2 Definition of design

Now, in the same design division that chose solution x_2 from those available in set O_X, a new department will undertake the detailed analysis of the selected solution. After a complex process comprised of different modeling, simulation and experimentation phases, it finds the functional relationship that links solution x_2 to product performance, which we will identify with the variable Y.[3] With this functional relationship, the detailed design that will be used later is created. In this usage, due to a set of imponderables, including ambient noise, variability due to the tolerances of the manufacturing processes, the different ways in which different operators work, and a long list of others, the performance achieved will vary slightly from the expected levels. Thus, variable Y, which quantifies real product performance, is not deterministic, and can take different values. Suppose that $\{y_1,...,y_n\}$ is the set whose elements are the labels identifying the different possibilities for the variable that defines performance,[4] i.e. $Y \in \{y_1,...,y_n\}$. Obviously, the existing constraints and the intensity of the random deviations mean that not all of the possibilities are equally probable: suppose that $(q_1,...,q_n)$ is the probability distribution for the variable Y. Finally, let us suppose that the customer is certain of the performance required to satisfy his needs. In other words, he knows exactly which of the labels in set $\{y_1,...,y_n\}$ is the one he needs; let's assume that this is y_3. We can conclude that the customer's degree of satisfaction with this product will not be complete because he receives $Y \in \{y_1, \ldots, y_n\}$ with the probability distribution $(q_1, \ldots, q_n)$, when he wanted $Y \in \{y_1, \ldots, y_n\}$ with the probability distribution (0,0,1,0,. . .,0). The satisfaction would have been maximum if $(q_1, \ldots, q_n) = (0,0,1,0, \ldots, 0)$ had been achieved.

The above discussion helps us to understand the design process as the process that seeks to transform a probability distribution that was initially similar to the uniform distribution into a deterministic distribution where all probabilities are null except one. The generalization of this reflection enables us to write the following definition of the design process:

Definition: *Design* is the formulation and execution of a plan to satisfy a need with the minimum degree of uncertainty.

The first part of this definition is identical to the one provided in Chapter 1. However, the second part explicitly indicates that the probability distribution changes as the design process progresses, and that the uncertainty associated with it will do the same. This aspect is closely related to the idea that 'to design is to define,' discussed earlier in Section 1.6. When the design process is executed correctly, the solution with the highest probability of being chosen coincides with the one that will provide the maximum satisfaction. On the other hand, at the beginning of the design process, the solution with the greatest probability of being chosen and the solution with the highest probability of success do not necessarily coincide.

2.3 Uncertainty

As we have seen, to design is to reduce the uncertainty about the satisfaction of a need through a decision-making process. With each decision, the designer must choose what he considers to be the best from a set of n possible paths. In this sense, each decision is a bifurcation. In the absence of any criteria for a choice, the designer will find them all equally feasible; in other words, equally probable. This suggests that information, uncertainty and probability distribution are related (Shannon, 1948). To explore this functional relationship, let us first consider the concept of uncertainty.

Suppose the solution set $O_X = \{x_1, x_2, \ldots, x_n\}$ exists, but is unknown to us. We don't know how many elements it contains, or what these elements are. Now suppose we are told that x_2 is an element of that set. Without any information on the other solutions, we cannot compare x_2 to the rest, and we cannot make any decision. However, if we are told that x_2 is an element, and that its probability of being chosen compared to the others is 1, we will not need any more information in order to choose it. All of the relevant information has been transmitted, even though we do not know how many other elements are available in the set, or what they are. In this sense, knowing that x_2 has a probability of 1 means we have zero uncertainty. On the other hand, if we were told that solution x_2, when compared to the other solutions, has a null probability of being chosen (i.e. is not an acceptable solution), we would know very little or nothing. We would know that x_2 must not be chosen, but that is the same as not knowing anything because we know nothing about the solution set. We don't know the number of possible solutions, or which is the correct one. Consequently, without a solution to pursue, we cannot progress in the decision-making process. In this case, we have infinite uncertainty because we know absolutely nothing about what to choose.

The above paragraph provides some clues for seeking a function that would enable us to take a probability and find the degree of uncertainty associated with it. For 0 probability, the uncertainty is maximum (let's say infinite), and for a probability of 1, it is zero. The function should also be sufficiently soft and monotonous, and additive for independent variables. One possible function is the logarithmic function.[5]

Definition: We define *uncertainty h* associated with probability p using the function:[6]

$$h : (0,1] \rightarrow [0,\infty),\ p \mapsto h(p) = -\log(p) \tag{2.1}$$

This definition is the basic foundation supporting the more complex concepts related to design. Entropy is an important part of these definitions, understood as the average value of the uncertainty of a distribution.[7] As we will see thanks to Gibbs' lemma, entropy decreases as the degree of certainty increases.

2.4 Entropy

The definition of the uncertainty of a probability, given in the previous section, enables us to define the uncertainty of a probability distribution as the average value of the uncertainty of each probability in the distribution. This average uncertainty is the entropy of the distribution.

Definition: *Entropy* H of probability distribution P is defined by the following function:[8]

$$H : \Delta(n) \to \mathbb{R},\ P \mapsto H(P) = \sum_{i=1}^{n} p_i h(p_i) \tag{2.2}$$

where $\Delta(n) = \{(p_1,p_2,...,p_n): p_i \in \mathbb{R},\ p_i \geq 0,\ \sum_{i=1}^{n} p_i = 1\}$ is the set of all possible probability distributions that can be assigned to a set containing n elements.

The number obtained by calculating the entropy is a characteristic of the distribution because it depends on it exclusively.[9] When a probability distribution is associated with a variable, it is also assigned an entropy as described in the following definition.

Definition: Let X be a variable that takes values in a finite set of n distinct elements $O_X = \{x_1,x_2,...,x_n\}$ referred to as the *alphabet* of variable X. Let p_i be the probability that variable X will adopt the value x_i (where $i = 1,2,...,n$). In other words, the probability distribution of variable X is $P_X = (p_1,...,p_n) \in \Delta(n)$. By definition, the *entropy of variable* X is $H(P_X)$.

Nomenclature: From now on, to avoid overloading the notation while writing the arguments for the entropy and information functions, we will eliminate the letter identifying the probability distribution, leaving only the letter identifying the variable. In other words, we will use the notation $H(X) = H(P_X)$. When necessary to specify several probability distributions associated with the same X variable, for example, $P_X \in \Delta(n)$ and $Q_X \in \Delta(n)$, we will add an argument to the variable, for example, $X(P)$ and $X(Q)$, or $X(t_1)$ and $X(t_2)$. A subscript in the variable will indicate a collection of distinct variables: X_i, where $i = 1,2,...,n$, will be a set of n distinct variables. (For example, $X_3(t_1)$ indicates the probability distribution associated with the variable X_3 in instant t_1.)

To find some of the most important properties of entropy, we would need the following lemma:

Gibbs' lemma: Let $a_i \geq 0$ and $b_i \geq 0$, where $i = 1,2,...,n$, be two sets of non-negative numbers, where $\sum_{i=1}^{n} a_i = \sum_{i=1}^{n} b_i$. These verify $-\sum_{i=1}^{n} a_i \log a_i \leq -\sum_{i=1}^{n} a_i \log b_i$, where the equality is true if, and only if, the two sets of numbers are identical; i.e. if $a_i = b_i$ for all of $i \in \{1,2,...,n\}$.

Proof: Let function $\varphi(x) = \ln x - x + 1$ be defined based on the real positive numbers. The first derivative is $\varphi'(x) = x^{-1} - 1$, and the second derivative is $\varphi''(x) = -x^{-2} \leq 0$ Because the second derivative is always negative, all of its extremes are maximum. Because it is continuous throughout its field of definition, it can only have one maximum. The first derivative is null in $x = 1$, where the function is $\varphi(1) = 0$. Under these conditions, we can argue that for any positive number, the function is never positive; in other words, $\varphi(x) \leq 0$ (where the equality is only valid if $x = 1$). In particular, $\varphi(a_i/b_i) \leq 0$ is met. Performing the operations, we obtain $-a_i \ln a_i + a_i \ln b_i \leq b_i - a_i$ (where the equality is only valid if $a_i = b_i$). By adding for every value of i on both sides of the inequality, taking into account the constraint described in the statement, the lemma is proven.[10]

This lemma is what makes entropy so important. First of all, if we invent two probability distributions that are the same size, the lemma states that the entropy of one of them is always bounded from above by a number that depends on both distributions. Note that this also occurs with any reordering of the distributions, and also when the first distribution is replaced by the second. In addition, the difference between the entropy of one distribution and the bound provided by this one and another is always positive if both distributions differ by at least two numbers.[11] Finally, using the following theorem, this lemma enables us to find an upper bound for the entropy in all of the distributions.

Theorem: The uniform probability distribution $(1/n,\ldots,1/n) \in \Delta(n)$ is the probability distribution in $\Delta(n)$ with maximum entropy.

Proof: To prove it, simply use Gibbs' lemma with $\sum_{i=1}^{n} a_i = 1$ and $b_i = 1/n$.

Furthermore, the entropy of the uniform distribution is $H((1/n,\ldots,1/n)) = \log n$. The following corollary is therefore verified.

Corollary: Given a probability distribution $P \in \Delta(n)$, its entropy always verifies:

$$H(P \in \Delta(n)) \leq \log n. \tag{2.3}$$

Another way of stating this is: any probability distribution with n elements containing at least one element with a probability of less than $1/n$ has an entropy lower than $\log n$. Therefore, the following property can be stated.

Property: The *upper extreme of the entropy* of a variable is the logarithm of the size of the alphabet.

The immediate conclusion applying to Design Science is that those situations with a large number of equally probable cases have a higher capacity to accumulate entropy. In other words, the greater the number of bifurcations in the decision tree, the higher the number of possible final cases and, consequently, the greater the entropy of the situation. The great usefulness of the concept of entropy in design is due to this theorem (and its corollary).

2.5 Joint entropy, conditioned entropy and relative entropy

In a design process, there is rarely a single variable. In this section, the concept of entropy is extended to situations with several variables.

Let $X \in \{x_1, \ldots, x_m\}$ and $Y \in \{y_1, \ldots, y_n\}$ be two variables that take their values in alphabets of size m and n, and whose probability distributions are $P = (p_1, \ldots, p_m)$ and $Q=(q_1, \ldots, q_n)$, respectively. Let $(X,Y) \in \{(x_1,y_1), \ldots, (x_1,y_n), \ldots, (x_m,y_1), \ldots, (x_m,y_n)\}$ be the variable that covers the Cartesian product $\{x_1, \ldots, x_m\} \times \{y_1, \ldots, y_n\}$ with the probability distribution $R = \{r_{11}, r_{12}, \ldots, r_{1n}, \ldots, r_{m1}, \ldots, r_{mn}\}$. These distributions meet the following constraints:

$$\sum_{i=1}^{n} p_i = \sum_{i=1}^{n} q_i = \sum_{i=1}^{m} \sum_{i=1}^{n} r_{ij} = 1 \qquad (2.4)$$

$$p_i = \sum_{j=1}^{n} r_{ij} \qquad (2.5)$$

$$q_j = \sum_{i=1}^{m} r_{ij} \qquad (2.6)$$

Definition: The *joint entropy* of variables X and Y is the entropy of variable (X,Y). In other words:

$$H(X,Y) = -\sum_{i=1}^{m} \sum_{j=1}^{n} r_{ij} \log r_{ij} \qquad (2.7)$$

Definition: The *entropy* of variable X *conditioned* by variable Y is:

$$H(X/Y) = -\sum_{i=1}^{m} \sum_{j=1}^{n} r_{ij} \log \frac{r_{ij}}{q_j} \qquad (2.8)$$

As we can see, given that r_{ij}/q_j is the probability of obtaining $X = x_i$, conditioned by the knowledge that $Y = y_j$ has been verified, the conditioned entropy is the average value of the uncertainty of the conditioned probability. Operating the logarithm, we can find the relationship between the different entropies.

$$H(X/Y) = -\sum_{i=1}^{m} \sum_{j=1}^{n} r_{ij} \log \frac{r_{ij}}{q_j} = H(X,Y) + \sum_{j=1}^{n} \log q_j \sum_{i=1}^{m} r_{ij} = H(X,Y) - H(Y) \qquad (2.9)$$

Definition: The entropy of variable X conditioned by a particular value y_j of variable Y is:

$$H(X/Y=y_j)=-\sum_{i=1}^{m}\frac{r_{ij}}{q_j}\log\frac{r_{ij}}{q_j} \tag{2.10}$$

With this definition, we can say that the entropy of variable X conditioned by variable Y is the average value of the entropies conditioned for each value of Y:

$$H(X/Y)=-\sum_{i=1}^{n} q_j\, H(X/Y=y_j) \tag{2.11}$$

Finally, using Gibbs' lemma, it is possible to define a relative entropy that we will call J.

Definition: The relative entropy of variables X and Y is:

$$J(X,Y)=\sum_{i=1}^{m} p_i \log\frac{p_i}{q_i} \tag{2.12}$$

Gibbs' lemma states that $J(X,Y) \geq 0$. The rest of the entropies defined are also non-negative, $H(X) \geq 0$, $H(X,Y) \geq 0$, $H(X/Y=y_j) \geq 0$ and $H(X/Y) \geq 0$, as the argument of the logarithm always belongs to the interval [0,1]. Because relative entropy is always non-negative, and only reaches a null value when the two distributions X and Y are identical,[12] it is very similar to a distance, but is not a distance.[13] For later proofs, the following two inequalities will be useful.

Property: The joint entropy of two variables is always greater than or equal to the entropy of one of them. Equality is obtained if X and Y are the same variable, i.e. if $r_{ij} = \delta_{ij}q_j$ where $\delta_{ij} = 0$ if $i \neq j$ and $\delta_{ij} = 1$ if $i = j$.

$$H(X,Y) \geq H(X) \tag{2.13}$$

Proof: This is immediate from (2.9) and $H(X/Y) \geq 0$. Equality is obtained by replacing $r_{ij} = \delta_{ij}q_j$ in (2.8).

Property: The entropy of a variable conditioned by another is always less than or equal to the entropy of the conditioned variable. Equality is obtained if, and only if, both variables are independent, i.e. if $r_{ij} = p_iq_j$. (A particular case is when variable Y is deterministic; in other words, when $r_{ij} = p_i\delta_{jk}$.)

$$H(X/Y) \leq H(X) \tag{2.14}$$

Proof: Simply operate using Gibbs' lemma with number sets r_{ij} and p_iq_j and relationship (2.5).

Chain rule: Conditioned entropy enables us to find the entropy of a set of variables. Let $X_i \in O_{X_i} = \{(x_i)_1,\ldots,(x_i)_{n(O_{X_i})}\}$, where $i = 1,2,\ldots,N$, be a set of variables that take their values in alphabets O_{X_i}, of size $n(O_{X_i})$, with probability distributions $P_i = ((p_i)_1, \ldots, (p_i)_{n(O_{X_i})})$. Starting from expression (2.9), we have:

$$H(X_1,X_2) = H(X_1) + H(X_2/X_1) \tag{2.15}$$

By replacing variable X_2 with (X_2,X_3):

$$\begin{aligned} H(X_1,X_2,X_3) &= H(X_1) + H(X_2/X_1,X_3/X_1) \\ &= H(X_1) + H(X_2/X_1) + H(X_3/(X_2,X_1)) \end{aligned} \tag{2.16}$$

If we proceed with any number of variables, we find the so-called chain rule, which enables us to find the joint entropy of an arbitrary number of variables:

$$H(X_1,X_2,\ldots,X_N) = \sum_{i=1}^{N} H(X_i/(X_{i-1},X_{i-2},\ldots,X_1)) \tag{2.17}$$

Chain rule (2.17) and properties (2.14) and (2.3) lead to the following inequality:

$$H(X_1,X_2,\ldots,X_N) \leq \sum_{i=1}^{N} H(X_i) \leq \sum_{i=1}^{N} \log n(O_{X_i}) \tag{2.18}$$

From this inequality, we learn that there are two ways to prevent a high value for the upper extreme of the entropy of a set of variables: 1) reducing the number of variables in the set, and 2) reducing the size of the alphabets.

2.6 Mutual information

If message Y, obtained from alphabet $\{y_1, \ldots, y_n\}$, of size n, decreases the uncertainty of variable X, obtained from alphabet $\{x_1, \ldots, x_m\}$, of size m, we would say that the message introduces information. Thus, for example, a message that reduces the uncertainty of a solution introduces information into the design process. The information is therefore associated with a difference of entropies.

Definition: The *mutual information* between two variables, X and Y, or the information that variable Y provides on variable X, is the entropy of X minus the entropy of X conditioned by Y:

$$I(X,Y) = H(X) - H(X/Y) \tag{2.19}$$

This expression can be interpreted as the result of a process[14] in which $H(X)$ is the initial entropy and $H(X/Y)$ is the final entropy after adding the knowledge of variable Y. If Y decreases the existing uncertainty of variable X, we would say that variable Y has added the amount of information $I(X,Y)$ to the design. The most important *properties* of mutual information are:

1. $I(X,Y) = \sum_{i=1}^{n}\sum_{j=1}^{m} r_{ij} \log \frac{r_{ij}}{p_i q_j}$ (2.20)

   Proof:

   $$I(X,Y) = H(X) - H(X/Y)$$
   $$= -\sum_{i=1}^{m} p_i \log p_i + \sum_{i=1}^{m}\sum_{j=1}^{n} r_{ij} \log \frac{r_{ij}}{q_j}$$
   $$= -\sum_{i=1}^{m}\sum_{j=1}^{n} r_{ij} \log p_i + \sum_{i=1}^{m}\sum_{j=1}^{n} r_{ij} \log \frac{r_{ij}}{q_j} \quad (2.21)$$
   $$= \sum_{i=1}^{m}\sum_{j=1}^{n} r_{ij} \log \frac{r_{ij}}{p_i q_j}$$

2. $I(X,X) = H(X)$ (2.22)

   Proof: Conditioned entropy is null when $r_{ij}/q_j = \delta_{ij}$; therefore, (2.19) concludes the proof.

3. $I(X,Y) \leq H(X)$ (2.23)

   Proof: Because the conditioned entropy is positive, this is immediately proven starting from (2.19).

4. $I(X,Y) \geq 0$ (2.24)

   Proof: Property 1 enables us to write mutual information as a relative entropy between the joint probability distribution r_{ij} of variable (X,Y) and distribution $p_i q_j$, obtained by considering variables X and Y independently. Because the relative entropy is non-negative, it is proven.

5. $I(X,Y) = 0$ if, and only if, X and Y are statistically independent.

   Proof: When the variables are statistically independent, we have $r_{ij} = p_i q_j$, which is replaced in expression (2.20) to conclude the first part of the proof. When $I = 0$, (2.21), this enables us to write $H(X/Y) = H(X)$. From here, we can conclude that the variables are independent.

6. $I(X,Y) = I(Y,X)$ (2.25)

   Proof: This is immediate when the distributions are permuted in Property 1.

7. $I(X,Y) = H(X) + H(Y) - H(X,Y)$ (2.26)

Proof:

$$I(X,Y) = H(X) - H(X/Y)$$

$$= H(X) + \sum_{i=1}^{m}\sum_{j=1}^{n} r_{ij} \log \frac{r_{ij}}{q_j} \quad (2.27)$$

$$= H(X) + H(Y) - H(X,Y)$$

Note that the bounds (2.23) of the mutual information can be specified more precisely:[15]

$$0 \leq I(X,Y) \leq \min\ (H(X),H(Y)) \quad (2.28)$$

This upper bound indicates that it is never possible to provide more information than the lowest entropy of the distributions solved. Furthermore, because the entropy of any distribution is always bound by the entropy of the uniform distribution, the following inequalities are always true:

$$0 \leq I(X,Y) \leq \min\ (H(X),H(Y)) \leq \min\ (\log m, \log n) \quad (2.29)$$

In some cases, this bound is what enables us to identify the information content of a design with the logarithm for the number of cases from which it is selected.[16] However, as we can see from the above bounds, this is only true when solving from the worst possible case, uniform distribution, to the best possible case, i.e. when we start with absolutely no criterion for selecting variable X, and finish with absolute certainty of variable X. Normally, the design process will not start with a complete initial lack of knowledge, nor will there be a perfect designer. However, it is true that the logarithm for the size of the smallest alphabet is an upper bound for the mutual information of two variables handled in real situations. Therefore, if the upper bound is decreased, the maximum 'amount' of consumable information, or the maximum required information, is decreased. From this perspective, the upper bound can be used for making design decisions.

2.7 Upper and lower bounds of mutual information. Information content of a variable

Inequalities (2.23) and (2.24) are the upper and lower bounds of the mutual information of two variables. To understand in what situations these limits are reached, we will analyze three extreme cases:

- **Case 1:** X and Y are independent random variables. In this case, the probability of (X,Y) is simply the product of probabilities, i.e. $r_{ij} = p_i q_j$. In this situation, we can see that $H(X/Y) = H(X)$. Consequently, variable Y does not contribute any information to the design, as it does not solve any of the uncertainty regarding the probability distribution of X.
- **Case 2:** X and Y are the same variable. Therefore, the probability of (X, Y) is $r_{ij} = \delta_{ij} q_j$, and the probability distributions of X and Y coincide, $p_i = q_i$; knowing what value Y has taken is the same as knowing what value X has taken. In this case, $H(X/Y = y_j) = 0$, $H(X/Y) = 0$ and, consequently, $I(X,Y) = H(X)$. In other words, the amount of information exactly matches the entropy of variable X. This result was already shown in (2.22), which states that the information a variable establishes about itself is its own entropy. In a design process, the entropy of a variable is not normally solved with the variable itself, but with another variable. Nonetheless, the information content of a variable is an upper bound of the mutual information of that variable and another one, $I(X,Y) \leq I(X,X)$. In information theory (Cover, 2006), $I(X,X)$ is usually called *self-information* or the *information content* of variable X.[17]
- **Case 3:** Variable Y is deterministic. In other words, it will definitely take the value y_k. The probability distribution is $q_j = \delta_{jk}$ and $r_{ij} = p_i \delta_{jk}$. In this case, $I(X,Y) = H(X) - H(X/Y = y_k) = 0$. The information introduced is null, as in Case 1. The information only appears when both variables are random and statistically dependent.

Case 1 shows a scenario in which an inexperienced designer does not know the causalities, and therefore chooses a Y variable that has nothing to do with the problem represented by X. In Case 2, an ideal designer finds the Y variable that completely solves the problem represented by X. The results of the Y variable found by this designer predict the results of X. In Case 3, the designer has chosen and fixed the value of a variable. No additional information can be expected from the other variables for this reason. Naturally, these are extreme cases. Real design situations would be somewhere in-between.

2.8 Process information

The mutual information of two variables calculates the decrease in the entropy of a first variable (the initial situation) when an activity causes the probability distribution of a second variable, which was not independent of the first variable, to become known (final situation). In general, we will adopt the following definition:

Definition: The *process* (activity) *information* is the difference in entropy between the initial and final states of that process (activity).

$$I = H_{INITIAL} - H_{FINAL} \tag{2.30}$$

A process can change the entropy because it modifies the probability distributions, changes the number of variables, or changes the size of the alphabets. Note that, unlike mutual information, process information can be negative. When the process information is positive (negative, respectively), the entropy decreases (increases, respectively) as a result of the process.[18] When all of the variables are independent, equality (2.17) leads to:

$$I = \sum_{i=1}^{N_{INITIAL}} H((X_{INITIAL})_i) - \sum_{i=1}^{N_{FINAL}} H((X_{FINAL})_i) \tag{2.31}$$

At this point, it is worth noting that the concept of information in a process is broader than proposed here. For example, Hyvärinen (1968) explained process information by making use of three levels: syntactic, semantic and pragmatic.

Syntactic level: This level is concerned with: 1) the number of possible symbols or labels available in the alphabets, 2) their probability distributions, and 3) the capacities of the communication or processing channels to efficiently and reliably process these labels. On this level, information theory is concerned with the transformation of the data, regardless of its meaning, importance or usefulness. However, the main characteristic of this level is uncertainty: before receiving a message, there must be more than one possible alternative for it. This is the type of information measured by the concept of mutual information.

Semantic level: This level is concerned with: 1) the recipient of the message, 2) the context in which the message is received, 3) the meaning of the symbols, and 4) the rules governing the formation of the language. For information to exist on this level, the recipient must be able to interpret and understand the message, i.e. have knowledge of the language. The difficulties on this level are normally resolved by means of conventionalisms or conventions between the sender and the recipient.

Pragmatic level: This level is concerned with: 1) the recipient of the message, 2) the context, and 3) the instant in which the message is received. For the recipient, the information must be relevant, or have value; it must serve to make a decision or start an activity. On this level, the dependence on the recipient and the context is even greater than on the semantic level. However, time has the greatest influence on the usefulness of the information. When irreversible decisions must be made at certain instants in time (deadlines), the usefulness of the information depends on how far in advance it was received. Once the deadline has passed, its usefulness will be low or null. To resolve the difficulties associated with this level, an appropriate approach is required, in keeping with a particular schedule and the proper execution of a set of activities and processes.

2.9 Spaces of definition: need–solution–response–satisfaction

The description provided in Chapter 1 establishes the mediation–invention–resolution–evaluation scheme as a fundamental characteristic of the design process. Mediation is basically concerned with lists of needs, invention with solutions, resolution with results, and evaluation with satisfaction. Needs, solutions, responses and satisfactions are defined by assigning certain values to certain variables. For this reason, in any design process we can identify four spaces of definition for variables: needs, solution, responses and satisfaction.

Space of definition for the needs: This is comprised of the alphabets that characterize the list of needs.[19]

Space of definition for the solution: This is comprised of the alphabets that completely describe the solution.[20]

Space of definition for the response: This is comprised of the set of alphabets that characterize the response for the solution. By response, we understand the actions or functionality of a solution. In this space, we obtain the results that must be compared to the needs.

Space of definition for the satisfaction: This is comprised of the set of alphabets that characterize the satisfaction of the solution. In this space, each point in the solution space is associated with a particular degree of acceptance based on the comparison between the results and needs. The simplest alphabet in this space is {*acceptance*, *rejection*}.

These spaces have the same structure, but different sizes. All variables and sets in the same space will be associated with the same letter. We will use W for the list of needs, X for the solution, Y for the response, and Z for the satisfaction. The structure of one of these spaces, which we will generically call space B so that we can refer to any of the four spaces interchangeably, is shown below.

Let $N(B)$ be the number of variables necessary to define space B. Let $B_1,B_2,...,B_{N(B)}$ be the variables that define that space. By definition, these variables are discrete and finite, i.e. they take their value from a finite set of elements called the *alphabet* of the variable. The alphabet of variable B_i is O_{B_i}. The alphabets of the different variables do not necessarily have the same number of elements. For this reason, we will refer to the number of elements comprising alphabet O_{B_i} as $n(O_{B_i})$. Each element of an alphabet is a *label*. The different labels $(b_i)_j$ that variable B_i can take are identified by the lower-case letter of the space and two subscripts, i and j. The first indicates the variable, and the second indicates the label. The relationship between a variable, its alphabet and its labels is $B_i \in O_{B_i} = \{(b_i)_1,...,(b_i)_{n(O_{B_i})}\}$ (where $i = 1,2,...,N(B)$). The set $\Omega_B = O_{B_1} \times O_{B_2} \times...\times O_{B_{N(B)}}$ obtained by finding the Cartesian product of all alphabets of B is the space of definition for variable $B = (B_1, B_2,...,B_{N(B)})$. An element of Ω_B is a point in the space of definition for variable B or, in more abbreviated form, of space B. The size of the space is $\left|\Omega_B\right| = \prod_{i=1}^{N(B)} n(O_{B_i})$. Because the spaces of definition are finite, the points in space Ω_B are numerable. We can therefore use order numbers to identify them in Ω_B. Set Ω_B is suitable for those spaces where the variables that define it can be independent, but there will be spaces where the vast majority of the points in Ω_B will be inaccessible.[21] For such cases, it is advisable to define set $\Xi_B = \{O_{B_1}, O_{B_2},...,O_{B_{N(B)}}\}$ and size $\left|\Xi_B\right| = \sum_{i=1}^{N(B)} n(O_{B_i})$.

The alphabets that generate the different spaces of definition are discrete and finite. For this reason, the discrete characteristics can be directly converted into a finite set of labels.[22] However, there are also characteristics described by continuous variables.[23] In these cases, a discretization process is required to obtain the alphabets.

Depending on space B, there can be a higher or lower number of continuous characteristics. Because this number depends on the space, we will call it $M(B)$. Obviously, it has to be $M(B) \leq N(B)$. By convention, alphabets originating from continuous characteristics are placed in the first $M(B)$ positions. Let $O_{B_1}, O_{B_2},...,O_{B_{M(B)}}, O_{B_{M(B)+1}}, O_{B_{M(B)+2}},...,O_{B_{N(B)}}$ be the alphabets that define space B. We will refer to set $\Xi|_B = \{O_{B_1}, O_{B_2},...,O_{B_{M(B)}}\}$ as the continuous space of definition for space B, and set $\Xi{:}_B = \{O_{B_{M(B)+1}}, O_{B_{M(B)+2}},...,O_{B_{N(B)}}\}$ as the discrete space of definition for space B. Alphabet $O_{B_i} = \{(b_i)_1,...,(b_i)_{n(O_{B_i})}\}$ belongs to $\Xi|_B$ when its labels identify intervals of $\mathbb{R}$ that constitute a finite partition of $\mathbb{R}$, i.e. when $(b_i)_1 \cup (b_i)_2 \cup...(b_i)_{n(O_{B_i})} = \mathbb{R}$ and $(b_i)_r \cap (b_i)_s = \emptyset$, where $r \neq s$ and $(b_i)_j \subset \mathbb{R}$ is an interval for any value of j.[24] In other words, the alphabets of $\Xi|_B$ were obtained by discretizing the set of real numbers $\mathbb{R}$ in each of the $p = M(B)$ dimensions of space $\mathbb{R}^p$. Assigning labels at intervals enables us to add needs quantified by continuous and discrete variables to the space of definition. The space will be the union of both sets of alphabets: $\Xi_B = \Xi|_B \cup \Xi:_B$.

2.9.1 Acceptance limits

The description of the list of needs will require a set with a number $N(W)$ of characteristics. Some of these characteristics will be discrete and finite, directly becoming the corresponding alphabets. Others will be continuous, giving rise to the continuous variables $M(W)$. The generation of the alphabet for these variables will be discussed in this section. We will see that in a design process, the acceptance limits enable us to define $\Xi|_W$. Let $(l_1,\ldots,l_p) \in \mathbb{R}^p$ be a set of $p = M(W) \leq N(W)$ real numbers quantifying continuous characteristics of the list of needs. It is the mediator's job to express the customer's initial motivation in terms of these numbers.[25] From the moment these p numbers reflect the customer's motivation, the solution proposed should not generate values significantly different from these. Otherwise, the customer would not have established them. Consequently, the customer's dissatisfaction will increase as the gap grows with respect to these values. It is also the mediator's job to express this dependency, and one possible way is by generating acceptance limits. For each number l_i, the mediator asks the customer for the maximum deviation he is willing to tolerate. As a result, the mediator will generate another two ordered lists of numbers for each, one indicating the lower acceptance limit, and the other indicating the upper acceptance limit. Let $\underline{l}_i < l_i < \overline{l}_i$, where $i = 1,2,\ldots,p$, be the $3p$ numbers that the mediator has specified as the definition of the need. For each i, these numbers divide the real line into three parts, $(-\infty, \underline{l}_i)$, $[\underline{l}_i, \overline{l}_i]$, and $(\overline{l}_i, \infty)$, where $[\underline{l}_i, \overline{l}_i]$ is the acceptance interval and the other two, the rejection intervals. With these intervals, the following alphabet can be built: {*rejection due to lack*, *acceptance*, *rejection due to excess*}. This is an alphabet with three elements, each associated with one of the above intervals.[26] There could be cases in which one of the numbers identifying the need has no clear upper or lower acceptance limit.[27] In this situation, the real line would be divided into two segments, $(-\infty, \underline{l}_i)$ and $[\underline{l}_i, \infty)$, where $[\underline{l}_i, \infty)$ is the acceptance interval. In general, each real line i of the p comprising space $\mathbb{R}^p$ can be divided into a number of segments $n(O_{W_i}) > 1$ which constitute the different elements $(w_i)_j$ of the alphabet in which variable $W_i \in \{(w_i)_1,\ldots,(w_i)_{n(O_{W_i})}\}$ takes values, where $i = 1,2,\ldots,p$ and $p = M(W)$. Each dimension of $\mathbb{R}^p$, i.e. each variable $W_i \in O_{Wi} \in \Xi|_W$, quantifies some continuous characteristic from the list of needs. Each partition in one of these dimensions establishes the alphabet $O_{W_i} = \{(w_i)_1, \ldots, (w_i)_{n(O_{W_i})}\}$, which is a gradation of the characteristic, and each element in it identifies a degree of acceptance by the customer.[28]

2.9.2 Tolerances

We will now see that in a design process, the tolerances define $\Xi|_X$. The procedure is similar to the one shown for acceptance margins, but we will work in the space of definition for the solution. A real space is established, $\mathbb{R}^q$ with $q = M(X) \leq N(X)$ dimensions. In each dimension i of $\mathbb{R}^q$, an alphabet $O_{X_i} = \{(x_i)_1, \ldots, (x_i)_{n(O_{X_i})}\}$ is established using the following procedure. Let m_i, where $i = 1,2,\ldots,q$, be a number covering one of the dimensions of $\mathbb{R}^q$. The partition of this dimension will therefore be established by setting a maximum limit $\bar{m}_i$, a minimum limit $\underline{m}_i$, and some tolerances. The limits indicate the extreme values that may be reached by a particular design or operation parameter. Intervals $(-\infty, \underline{m}_i)$ and $(\bar{m}_i, \infty)$ therefore have a null probability of being chosen. Interval $[\underline{m}_i, \bar{m}_i]$ is in turn divided into intervals depending on the desired precision for the design parameter. For example, for a constant tolerance of $\pm\delta_i$ throughout the interval, the partition would be $[\underline{m}_i, \underline{m}_i + 2\delta_i]$, $[\underline{m}_i + 2\delta_i, \underline{m}_i + 4\delta_i]$, $[\underline{m}_i + 4\delta_i, \underline{m}_i + 6\delta_i]$, $\ldots$, $[\bar{m}_i - 2\delta_i, \bar{m}_i]$. It is part of the designer's job to set these intervals, which are the alphabets for each dimension. While the partitions in the space of definition for needs normally have 2 or 3 elements, the partitions in the space of definition for solutions have many more because the operation and design parameters can vary by amounts in the order of their own value, or may even shift by several orders of magnitude, while the error required of them is much less than their own value.

2.9.3 Transfer functions

Transfer functions[29] are the nexus between the space of definition for solutions and the space of definition for responses. The transfer function for a solution is a function that relates the operation and design parameters to the response and verifies these conditions: 1) it is a function $f: C \to \mathbb{R}^r$ where $C = [\underline{m}_1, \bar{m}_1] \times [\underline{m}_2, \bar{m}_2] \times \ldots \times [\underline{m}_q, \bar{m}_q] \subset \mathbb{R}^q$ and 2) it is a continuous function in C. The continuity of the function makes it possible to establish the following property.

If $I_i \subset [\underline{m}_i, \overline{m}_i] \subset \mathbb{R}$ is a set of intervals, the set $A = I_1 \times I_2 \times ... \times I_q \subset C \subset \mathbb{R}^q$, obtained by finding the Cartesian product of those intervals, has as its image the set $f_1(A) \times f_2(A) \times ... \times f_r(A) \subset \mathbb{R}^r$, where $f_i(A) \subset \mathbb{R}$ is an interval corresponding to the image of the *i*th coordinated function of the transfer function.[30]

A transfer function is defined as a function that takes values in $\mathbb{R}^q$ and returns them in $\mathbb{R}^r$. For this reason, it is a function that relates points from $\Omega|_X$, where $q = M(X) \leq N(X)$, to points from $\Omega|_Y$, where $r = M(Y) \leq N(Y)$. However, the form of a transfer function can change depending on the value adopted by the discrete variables defining the solution. For this reason, transfer functions are $f_a : C \rightarrow \mathbb{R}^p$ where $C \subset \mathbb{R}^q$, and where the subscript $a \in \Omega{:}_X$ indicates a possible combination of the discrete labels. This involves assuming that for each element a in set $\Omega{:}_X$, there is a transfer function f_a with the coordinated functions $(f_a)_1, ..., (f_a)_q$.

2.9.4 Response margins

We will see that the transfer functions establish the alphabets in the space of definition for the response. Indeed, we will now build the sets $A_{i_1 i_2 ... i_q} = (x_1)_{i_1} \times (x_2)_{i_2} \times ... \times (x_q)_{i_q} \subset C \subset \mathbb{R}^q$, where $i_j = 1,2,...,n(O_{W_j})$, based on the intervals $(x_i)_j$ that were obtained while establishing the tolerances. Thanks to the continuity of the coordinated functions of the transfer function, the image of these sets is an interval in each of the dimensions of $\mathbb{R}^r$. One of the intervals associated with the coordinated function $(y_j)_{i_1 i_2 ... i_p, a} = (f_a)_j(A_{i_1 i_2 ... i_p}) \subset \mathbb{R}$ will be called j. The transfer function property proven earlier enables us to build the alphabets $\{(y_j)_{i_1 i_2 ... i_p, a} : i_k = 1,2,...,n(O_{X_k}) \wedge a \in \Omega{:}_X\}$ in $\Omega|_Y$, where $j = 1,2,...,r$. As we can see, the label set in the alphabets in $\Omega|_Y$ depends on the value taken by the discrete variables defining the solution. Consequently, the above set can be written as: $\{(y_j)_{i_1 i_2 ... i_{N(X)}} : i_k - 1,2,...,n(O_{X_k})\}$, where $j = 1,2,...,M(Y)$. If the discrete variables of $\Omega{:}_Y$ are also added, we have $\{(y_j)_{i_1 i_2 ... i_{N(X)}} : i_k = 1,2,...,n(O_{X_k})\}$, where $j = 1,2,...,N(Y)$ or, in condensed notation: $O_{Y_j} = \{(y_j)_a : a \in \Omega_X\}$, where $j = 1,2,...,N(Y)$. We refer to each of the elements in alphabets $O_{Y_i} = \{(y_i)_1, ..., (y_i)_{n(O_{Y_i})}\}$ as the response margin. Note that the size of all alphabets in Ω_Y is the same: $n(O_{Y_i}) = |\Omega_X|$, where $i = 1,2,...,N(Y)$.

The relationship between acceptance limits, tolerances, transfer functions and response margins is represented graphically in Figure 2.1

Figure 2.1 **Relationship between the different spaces of definition and the transfer function.**

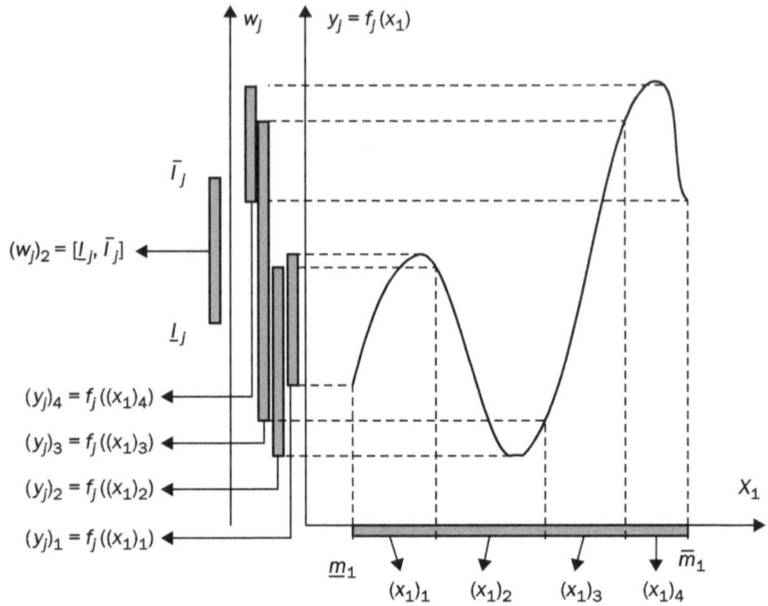

Note: The axis of abscissas represents a continuous characteristic, x_1, of the space of definition for the solutions. The discretization imposed by the tolerances on the interval between the upper and lower limits has been reduced, for the sake of simplicity, to four intervals producing four labels,$(x_1)_1$,. . .,$(x_1)_4$. On the ordinate axis, for a need depending only on x_1 both the acceptance interval and the response labels are represented.

for a case in which the space of definition for the solutions has a single alphabet with four labels obtained by discretizing a continuous variable.

2.9.5 Satisfaction

Each element $a \in \Omega_X$, which we call article, generates a response that must be compared to the list of needs. This comparison results in the rejection or acceptance of *a*. Thus, in the space of definition for satisfaction, each article is assigned the variable Z_a, which can take two values: acceptance or rejection. For this reason, all alphabets of Ξ_Z are identical: $O_{Z_i} = \{(z_i)_1,(z_i)_2\} = \{acceptance,\ rejection\}$, where $i = 1,2,\ldots,N(Z)$ is $N(Z) = |\Omega_X|$.

Defining the degree of satisfaction requires comparing the response to the needs that the solutions should have satisfied. Therefore, the number of alphabets in Ξ_W and Ξ_Y must coincide. From this, we deduce that $N(Y) = N(W)$. In addition, the alphabet O_{Y_i} must refer to the response achieved to satisfy need O_{W_i}, and not another. In general, however, the number of elements in alphabets Ξ_W and Ξ_Y will not coincide: $n(O_{Y_i}) \neq n(O_{W_i})$.

2.9.6 Encoding

As we have seen, when the variable l_j, associated with the space of definition for the needs, has clear lower and upper acceptance limits, $\underline{l}_j$ and $\overline{l}_j$, respectively, the intervals $(-\infty, \underline{l}_j)$, $[\underline{l}_j, \overline{l}_j]$, and $(\overline{l}_j, \infty)$ define the alphabet {*rejection due to lack*, *acceptance*, *rejection due to excess*}. If this occurs for all variables of $\Xi|_W$, the space can be encoded so that the alphabet {*rejection due to lack*, *acceptance*, *rejection due to excess*}, common to all of them, has each of its elements associated with the same set of three intervals. It is advisable to set these intervals to $(-\infty, -1)$, $[-1,+1]$, and $(+1,\infty)$. If this definition is adopted for the needs, the necessary comparison of the response to the need will require redoing the response margins using the following expression:

$$(y_j)_{i_1 i_2 \ldots i_p, a} = \frac{2(f_a)_j(A_{i_1 i_2 \ldots i_p}) - \overline{l}_j - \underline{l}_j}{\overline{l}_j - \underline{l}_j} \tag{2.32}$$

This expression encodes all lower acceptance limits as -1, all upper limits as $+1$, and the center of the acceptance interval as 0.[31] Thus, we can say that the spaces of definition for the response and the need are encoded to -1 and $+1$, where it is known that the probability of acceptance is 1.0 for values between -1 and $+1$, and null for values under -1 or over $+1$.

2.10 Degree of satisfaction

In this section, we will see how the degree of satisfaction generated by a point in the space of definition for solutions can be found as the probability that the response will satisfy the needs, i.e. as the probability of success.

Let $P_W \in \Delta(|\Omega_W|)$ be the probability distribution that assigns a probability $p_e = \Pr|_{\Omega_W}(e)$ of being accepted to each element $e \in \Omega_W$.[32] Let $P_X \in \Delta(|\Omega_X|)$ be the probability distribution that assigns a probability $p_a = \Pr|_{\Omega_X}(a)$ of being accepted to each element $a \in \Omega_X$.[33] The transfer functions assign a label set $(y_i)_a$, where $i = 1,2,\ldots,N(Y)$, to each article $a \in \Omega_X$. Let $P_Y \in \Delta(|\Omega_X|)$ be the probability distribution that assigns the probability of article $a \in \Omega_X$ being accepted to label $(y_i)_a$. All Ξ_Y alphabets are therefore associated with the same probability distribution, inherited from the space of definition for solutions. Furthermore, $Y = (Y_1,\ldots,Y_{N(Y)}) \in \Omega_Y$ has a probability distribution given by $\Pr|_{\Omega_Y}(((y_1)_{i_1},\ldots,(y_{N(Y)})_{i_{N(Y)}})) = \delta_{i_1\cdots i_{N(Y)}} \Pr|_{\Omega_X}(i_1)$, where $\delta_{i_1\cdots i_{N(Y)}}$ is 1 if all of its subscripts are equal, and 0 otherwise. Spaces Ω_X and Ω_Y therefore have the same entropy.[34]

In space Z, each element of Ω_X is assigned the variable Z_i, where $i = 1,2,\ldots,N(Z)$ is $N(Z) = |\Omega_X|$. The probability distribution of variable Z_a is called $P_{Z_a} \in \Delta(2)$. In this space, each article $a \in \Omega_x$ is assigned the variable Z_a with the probability distribution $(p_a,(1-p_a))$ which reflects the probability of being accepted or rejected.[35] The value of this probability depends on the needs and the response to the solution, i.e. $p_a = \Pr|_{\Omega_W,\Omega_X}(a)$. The calculation of these probabilities requires the following definitions.

Definition: Given the interval $I \subset \mathbb{R}$, the lower extreme of I is called $\inf(I)$ and the upper extreme of that interval is called $\sup(I)$.

Definition: Given two intervals, I_1 and I_2, the *common range* is:

$$\mathrm{cr}(I_1,I_2) = \begin{cases} 0 & \text{if } \Delta_{21}\Delta_{12} \le 0 \\ \min(\Delta_{21},\Delta_{12},\Delta_{11},\Delta_{22}) & \text{otherwise} \end{cases}$$

$$\Delta_{21} = \sup(I_2) - \inf(I_1) \quad ; \quad \Delta_{12} = \sup(I_1) - \inf(I_2)$$
$$\Delta_{11} = \sup(I_1) - \inf(I_1) \quad ; \quad \Delta_{22} = \sup(I_2) - \inf(I_2) \tag{2.33}$$

This function is shown in Figure 2.2.

Properties:

1. $\mathrm{cr}(I_1,I_2) = \mathrm{cr}(I_2,I_1)$ (2.34)

   Proof. Simply switch the subscripts in (2.33).

Figure 2.2 Common range of two intervals

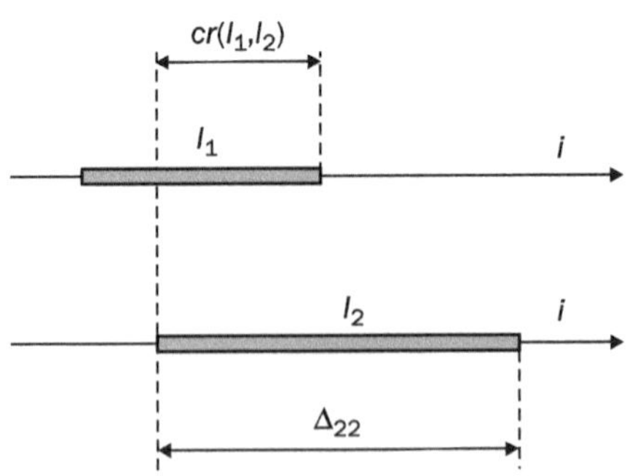

2. $0 \leq cr(I_1,I_2) \leq \min(\Delta_{11},\Delta_{22}) \leq \Delta_{22}$ (2.35)

   Proof. The arguments eliminated from the function are either less than the arguments kept, in which case the inequality is true, or greater than or equal to the ones kept, in which case the equality is true.

3. $cr(I_1,I_2)$ is maximum when $I_1 \subseteq I_2$ or $I_2 \subseteq I_1$. (A particular case is when the intervals are centered.)

   Proof. Thanks to Property 1, we can impose $I_1 \subseteq I_2$ with no loss of generality. This condition imposes $\inf(I_2) \leq \inf(I_1)$ and $\sup(I_1) \leq \sup(I_2)$. From these inequalities, we deduce $\Delta_{11} \leq \Delta_{22}$, $\Delta_{11} \leq \Delta_{12}$ and $\Delta_{11} \leq \Delta_{21}$. Because $\Delta_{11} > 0$, (2.33) yields $cr(I_1,I_2) = \Delta_{11}$. If a positive displacement is imposed on interval I_1, in such a way that $I_1 \subseteq I_2$ is not verified, we will obtain $\inf(I_2) < \inf(I_1)$ and $\sup(I_1) > \sup(I_2)$. From here, we can deduce $\Delta_{21} < \Delta_{11} < \Delta_{12}$. The function will therefore be $cr(I_1,I_2) = \max(0,\Delta_{21}) < \Delta_{11}$, which proves the property.

Definition: Given two intervals, $I_1, I_2 \subset \mathbb{R}$, the probability that a point of interval I_2 belongs to interval I_1 is the *interval probability* of I_1 and I_2:[36]

$$ip(I_1,I_2) = \frac{cr(I_1,I_2)}{\Delta_{22}} \tag{2.36}$$

Properties:

1. Not symmetrical: $ip(I_1,I_2) \neq ip(I_2,I_1)$. (2.37)
2. $0 \leq ip(I_1,I_2) \leq 1$. Proven from (2.35).
3. $ip(I_1,I_2)$ is maximum when $I_1 \subseteq I_2$ or $I_2 \subseteq I_1$. (A particular case is when the intervals are centered.) This is proven by Property 3 of function *cr*.

Definition: Given $O_{W_i} \in \Xi|_W$ and $O_{Y_i} \in \Xi|_Y$, where $i = 1,2,\ldots,M(Y)$, the probability of success of article a for satisfying need W_i is defined as $ip((w_i)_j,(y_i)_a)$, where $(w_i)_j \in O_{W_i} = \{(w_i)_1,\ldots,(w_i)_{n(O_{W_i})}\}$ is the label that characterizes the need, and $(y_i)_a \in O_{Y_i} = \{(y_i)_a : a \in \Omega_X\}$ is the label that characterizes the response.[37]

Definition: Given two elements, a and b, the discrete probability of a and b is:

$$dp(a,b) = \begin{cases} 0 & \text{if } a \neq b \\ 1 & \text{if } a = b \end{cases} \tag{2.38}$$

Definition: Given $O_{W_i} \in \Xi:_W$ and $O_{Y_i} \in \Xi:_Y$, where $i = M(Y),\ldots,N(Y)$, the probability of success of article a for satisfying need W_i is defined as the discrete probability of $(w_i)_j$ and $(y_i)_a$, i.e. as $\mathrm{dp}((w_i)_j, (y_i)_a)$, where $(w_i)_j \in \mathrm{O}_{W_i} = \{(w_i)_1, \ldots, (w_i)_{n(\mathrm{O}_{W_i})}\}$ is the label that characterizes the need and $(y_i)_a \in \mathrm{O}_{Y_i} = \{(y_i)_a : a \in \Omega_X\}$ is the label that characterizes the response.

Satisfaction: Given $\{(w_i)_1, \ldots, (w_i)_{n(\mathrm{O}_{W_i})}\}$ and $\{(y_i)_a : a \in \Omega_X\}$, where $i = 1, 2, \ldots, N(Y)$, the *probability of success* of article $a \in \Omega_X$ is defined as:[38]

$$\begin{aligned}\Pr|_{\Omega_W,\Omega_X}(a) = {} & \sum_{i=1}^{M(W)} \sum_{j=1}^{n(\mathrm{O}_{W_i})} \Pr|_{\Omega_W}\big((w_i)_j\big) \Pr|_{\Omega_X}(a) \mathrm{ip}\big((w_i)_j,(y_i)_a\big) \\ & + \sum_{i=M(W)}^{N(W)} \sum_{j=1}^{n(\mathrm{O}_{W_i})} \Pr|_{\Omega_W}\big((w_i)_j\big) \Pr|_{\Omega_X}(a) \mathrm{dp}\big((w_i)_j,(y_i)_a\big)\end{aligned} \tag{2.39}$$

The success of each article can serve to select it. To do so, we assign a new probability distribution to space Ω_X, obtained by normalizing the probabilities of each article for the entire space:

$$\Pr|_{\Omega_X}(a) = \frac{\Pr|_{\Omega_W,\Omega_X}(a)}{\sum_{a \in \Omega_X} \Pr|_{\Omega_W,\Omega_X}(a)} \tag{2.40}$$

2.10.1 Stop criterion

As shown, the acceptance of a solution requires the accomplishment of two conditions: 1) the probability of success (Eq. 2.39) must be maximum and greater than a minimum tolerable value, and 2) the probability of selecting that solution (Eq. 2.40) must be one. The fulfillment of both conditions leads to the maximum *probability of acceptance* for that solution.

2.11 Conceptual and detailed design

One of the objectives of the design process is to choose the appropriate value for a certain parameter. When a particular element of set Ω_X (that is, a particular combination of values for the different parameters) does not lead to an appropriate response, it must be ruled out. The process of eliminating elements leads to a refinement of the solution. As mentioned in Section 2.1, one way of carrying out this process of elimination is to replace the probability distribution $P_X(t_1)$ assigned to set Ω_X with another $P_X(t_2)$, in which a null probability is imposed on the element to be eliminated. For this reason, a solution can have different degrees of definition. The definition of the solution increases as its average uncertainty is reduced; in other words, as the number of zeros (or values close to zero) increases in the probability distributions of the alphabets. Because the entropy of a set does not change if the elements with null probability are eliminated from it, we can extract from any set the subset with the smallest number of elements that maintain the entropy of the configuration. If the unviable elements are eliminated from a solution, we obtain a *conceptual design*. We can describe a *detailed design* as a subset of a conceptual design, obtained by eliminating the elements with an inappropriate response; an *article* is an element of a detailed design.[39] Within a particular conceptual design (CD), there can be several different detailed designs (DD) and, within these, different articles (a). The relationship between them is:[40] $a \in DD \subset DC \subset \Omega_X$.

2.12 Operators. Necessary, generated and available information

The four spaces of definition can be associated with four operators. The *mediation operator* is responsible for turning the motivation into a list of needs, and creating the space of definition Ξ_W for the needs. The *synthesis operator* takes space Ξ_W and creates a possible solution that it turns into the space of definition Ξ_X for the solutions. The *calculation operator* takes space Ω_X, generates all of the transfer functions associated with it, and uses these to create the space of definition for the response Ξ_Y. Finally, the *evaluation operator* takes space Ξ_Y and compares it to space Ξ_W to generate space Ξ_Z.

Each of these spaces has an associated probability distribution. Thus, each element in the space has a probability of being accepted as a final decision. The corresponding operator creates the space along with its probability distribution. For example, in space Ω_W, the mediation operator will mark the labels that will be rejected by the customer with a zero, and the ones that will be accepted with a one. In space Ω_X during a conceptual design phase, a uniform distribution will be assigned to all labels identifying positions of a design parameter not detailed in that phase.[41] During the detailed design phase, all design labels eliminated in the conceptual design phase will be marked with zero probability because they were outside the requested ranges. The calculation operator obtains the response margins, and assigns them to Ξ_Y based on the transfer functions and distributions in Ω_X. The evaluation operator includes the stop or continue criterion. The evaluator compares the elements in space

Ξ_W to the elements in space Ξ_Y and generates space Ξ_Z. By selecting the best responses, it generates a new distribution associated with Ω_X, which takes into account how close or far an article comes to producing a result accepted by the customer. In this way, part of the solutions in Ω_X tend to reduce their probability, and therefore to disappear. If the most probable result has a relatively low probability of acceptance with the new probability distribution, the design process must begin again. However, in the new reformulation of the problem, the mediation operator will have the information generated in the previous step, indicating which solutions died and which were the most satisfactory. With this new information, the mediator makes a new list of needs and, consequently, a new Ω_W space, which will prompt the synthesis operator to generate a new Ω_X space, and so on until the stop criterion is satisfied. This is shown in Figure 2.3.

Figure 2.3 **Spaces of definition and operators**

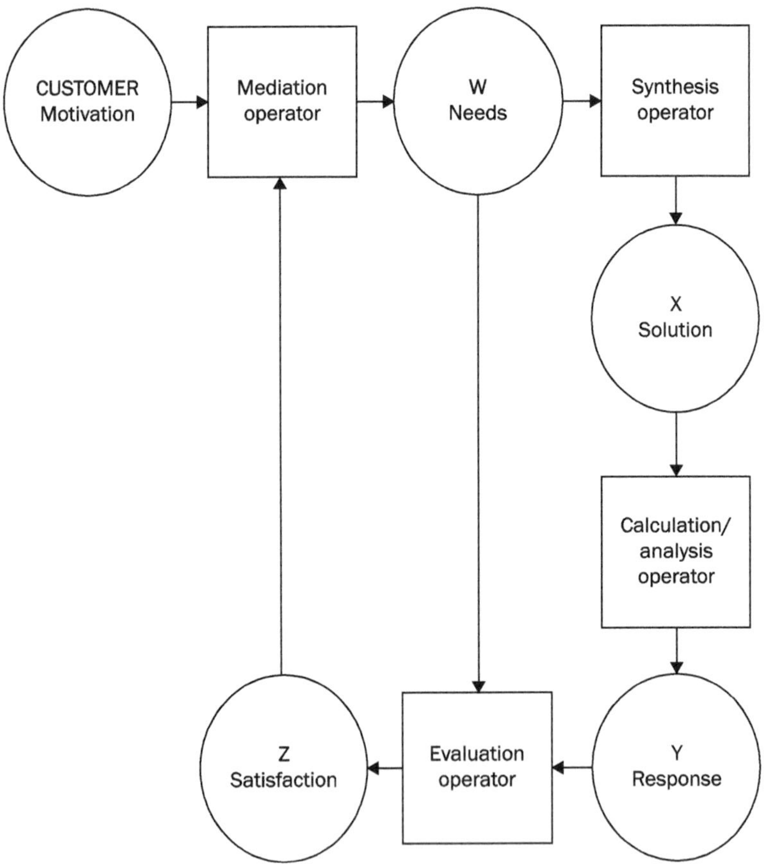

Because the evaluator must also perform calculations, we will refer to the operator that combines calculation and evaluation operators into one as the *analysis operator*. The analysis operator chooses those results that are most satisfactory for the customer, and uses them to change the probability distributions in Ω_X and Ω_Y, thus reducing the entropy. The analysis operator is the only agent responsible for reducing entropy. The mediation and synthesis operators, on the other hand, increase it by creating the three spaces of definition.[42]

When an operator generates the space of definition associated with it, it generates entropy. The greater the number and size of the alphabets, the greater this entropy. On the other hand, when an operator reduces the uncertainty associated with a space, it eliminates entropy. These modifications of the entropy are incoming or outgoing information. Thus, during the design process, the operators work with various types of information.

Starting information: This is associated with the motivation.

Necessary information: This is associated with the entropy of the spaces, and therefore their size. It is called necessary information because the process concludes when the uncertainty in those spaces decreases sufficiently due to the addition of this amount of information.[43]

Generated information: This is associated with the reduction of entropy achieved. Obviously, the generated information must be as close to the necessary information as possible.

Available information: This is associated with the very structure of the operators, and forms part of their know-how. Thanks to this information, the operators know how to create the spaces using both the starting and the generated information. This includes knowledge of how to devise, invent or conceive solutions (a creative process); knowledge of how to establish transfer functions, calculate, compare, etc. (a technological process); and knowledge of where to find that information (a documentary process). Part of the information available to the operators consists of the conclusions that can be reached from the definitions and relationships described in this chapter, some of which are discussed in the following section.

2.13 First statements

The statements discussed in this section are obtained from the approaches followed in the previous sections. When these statements are included as part of the available information, the operators responsible for carrying out the design process will modify the way they operate, and the design process will inevitably be different.

Statement #1: The lower the upper extreme of the entropy of the spaces in question, the less necessary information (and generated information) there will be.

Proof and explanation: We have seen (Eq. 2.28) that it is never possible to provide more information than the lowest entropy of the distributions solved. Therefore, no matter how many resources have been consumed, the generated information will never exceed the necessary information.[44] On the other hand, the necessary information is limited by the entropy of the uniform distribution with the smallest size (Eq. 2.29), which coincides with the upper extreme of the entropy (Eq. 2.3). If we start from an initial situation of maximum uncertainty, the statement is true. Therefore, if the upper bound is decreased, the maximum 'amount' of 'required' information, or maximum 'consumable' information, is decreased, as indicated at the end of Section 2.6.

Statement #2: For a set of independent variables with a common alphabet of more than four variables, the decrease in relative terms of the upper extreme of the entropy is greater when a certain percentage of the variables is eliminated than when the same percentage of elements in the alphabet are eliminated. In other words, to decrease the entropy, it is more profitable to decrease, in relative terms, the number of variables than to decrease the size of the alphabet.

Proof and explanation: For a set of independent variables uniformly distributed among a common alphabet, the upper extreme of the entropy is linear to the number of variables, and logarithmic to the size of the alphabet (when uniformly distributed, the equalities in 2.18 are valid). Hence, it follows that $H = k\,N \ln n$. If we decrease N by ΔN and n by Δn, then $\Delta_N H = -k \ln n\, \Delta N$ and $\Delta_n H = k\,N \ln(n - \Delta n) - k\,N \ln n$. Hence, $-\Delta_N H/H = \Delta N/N$ and $\Delta_n H/H = \ln(1 - \Delta n/n)/\ln n$. Given that $-n \ln(1 - \Delta n/n)/\Delta n/\ln n < 1$ for $n > 4$ and $n - \Delta n \geq 2$ (this is also verified for $n = 4$ and $\Delta n = 1$), $-\Delta_n H/H < \Delta n/n$ is proven. Therefore, even though the entropy is reduced when we reduce N and when we reduce n, it is preferable to keep the number of variables limited rather than the number of labels in the variables.

Statement #3: The upper extreme of the entropy of a design process is decreased when the upper extremes of the entropy of the spaces of definition for the needs and the solution are decreased.

Proof and explanation: As we saw in Section 2.9, the sizes of the different spaces of definition verify the relationship $|\Omega_X| = N(Z) = \frac{|\Xi_Z|}{2} \leq |\Xi_Y| = N(W)|\Omega_X|$. On the other hand, the entropies of the different spaces of definition verify the relationship $H(X) = H(Y) \leq \log |\Omega_X|$ and $H(Z) \leq |\Omega_X| \log 2$. Therefore, the reduction in the size of the space of definition for the solution reduces the upper extreme of the entropy of that space, and of spaces Y and Z. The reduction in the upper extreme of the entropy of the space for needs requires an independent reduction of its size $|\Omega_W|$.

Statement #4: The stricter the tolerances, the higher the upper extreme of the entropy.

Proof and explanation: The greater the quotient $(\overline{m}_i - \underline{m}_i)/2/\delta_i$, the higher the number of labels created in alphabet O_{X_i}. The statement can be proven immediately because the size of the alphabets, $n(O_{Xi})$, increases.

Statement #5: Any of these actions will help increase the probability of acceptance: 1) Centering the intervals of $\Xi|_Y$ with respect to the intervals of $\Xi|_W$ by displacing the intervals in $\Xi|_X$, 2) Increasing the acceptance limits in $\Xi|_W$, 3) Decreasing the tolerances in $\Xi|_X$, and 4) Decreasing the slopes[45] in the transfer function.

Proof and explanation: The probability of acceptance is given by (2.39) and (2.40). To prove this, 1) simply note that (2.39) and (2.40) are maximum when function *ip* is maximum, which occurs when the intervals are centered, thanks to Property 3) of that function. Increasing the acceptance limits favors the situation $(y_i)_a \subseteq (w_i)_j$, so that function *ip* either does not change or increases. This proves 2) decreasing the tolerances in $\Xi|_X$ decreases the length of intervals $(y_i)_a$, favoring the condition $(y_i)_a \subseteq (w_i)_j$ and reducing the denominator in (2.36). This proves 3) when the absolute value of the slopes in the transfer function are decreased, the length of intervals $(y_i)_a$ is also reduced, proving part 4) of the statement.

Statement #6: The starting information and the available information must be maximum for the entropy to be minimum.

Proof and explanation: The entropy in each of the spaces of definition is characterized by the variables W_i, X_j and Y_k, where $i = 1, 2, \ldots, N(W)$, $j = 1,2, \ldots, N(X)$ and $k = 1,2, \ldots, N(Y)$. The greater the starting information and the available information, the greater the number of information records (i.e. known variables and distributions) and, consequently, the higher the probability of finding dependencies between the information records (IR) and variables W_i, X_j and Y_k. The proof of this statement comes from property (2.14): $H(W, X, Y/IR) \leq H(W, X, Y)$.

Statement #7: The greatest reduction in the entropy is obtained when the design team is comprised of a multidisciplinary group of experts.[46]

Proof and explanation: The greater the number of different experts, the greater the available information. The proof is concluded by making use of the previous statement.

2.14 Notes

1. In general, success can be understood as the correct satisfaction of the needs, or the correct response to the motivation. In this same chapter, Equation (2.39) provides an initial approach to such probability. In the context of Axiomatic Design, which will be introduced in the next chapter, the probability of success is the probability that the solution conceived will meet the functional requirements and specified constraints (see Eq. (3.1)).
2. Whether or not the information is useful depends greatly on the 'level of pragmatism'. See Section 2.8.
3. Product performance is normally identified with several variables, Y_1, Y_2, . . ., Y_3. (An engine, for example, can be identified by its power, fuel consumption, useful life, etc.) In turn, with these n one-dimensional variables, a variable with n dimensions can be constructed $Y = (Y_1, \ldots, Y_n)$, which completely specifies the response. This generalization is addressed in greater detail in Section 2.9, in the discussion of the space of definition for the response.
4. In general, each label specifies a certain level (or range) of performance. The definition of each label depends on the particular design problem. For example, for a gasoline engine, the label y_3 could specify power greater than or equal to 100 kW, y_2 could indicate power between 50 and 100 kW, and y_1 could specify power less than or equal to 50 kW. If the performance were to include fuel consumption in addition to power, the label y_3 could indicate, for example, consumption under 210 g/kWh and power over 100 kW. For a customer, if the definitions of the labels are adequate, there will only be one label with the maximum acceptance.
5. As we will see in the next section, the uniqueness theorem for entropy states that, under certain hypotheses, the logarithmic function is the only one available for measuring uncertainty.
6. The logarithm can have any base. Thus, if the base is defined as 2, the unit of measure will be BIT. If the natural base is used, it will be NAT, and if the decimal base is used, it will be DIT.
7. To avoid confusion later while using the axioms and their conclusions, both the Information and the related concepts must be defined quite rigorously.
8. In $p = 0$, the value of $-p\log p$ is not defined; $0\log 0 = 0$ is adopted as a convention in order to maintain the continuity of the function. When the logarithm is taken to base two, it is usually referred to as Shannon's entropy. Except for one constant, it also coincides with Boltzmann's entropy. Boltzmann's entropy has physical units (Joules/Kelvin), while the entropy defined here is adimensional. (Although not necessary, it is useful to indicate the logarithm base by adding BIT, NAT or DIT.) The difference between the two is the base taken for the logarithm and Boltzmann's constant: $k=1.38 \cdot 10^{-23}$ J K^{-1}.
9. *Uniqueness theorem.* Given the following distributions, $U = (1/n, \ldots, 1/n) \in \Delta(n)$, $P = (p_1, \ldots, p_n) \in \Delta(n)$, $Q = (p_1, \ldots, p_n, 0) \in \Delta(n + 1)$, and $R = (p_1, \ldots, p_{j-1}, p_j + p_k, p_{j+1}, \ldots, p_{k-1}, 0, p_{k+1}, \ldots, p_n) \in \Delta(n)$, the uniqueness theorem states that the only function that is symmetrical and continuous in all of its variables, which verifies 1) $H(P) \leq H(U)$, 2) $H(P) = H(Q)$, and 3) $H(P) =$

$H(R) + (p_j + p_k)H(p_j/(p_j + p_k), p_k/(p_j + p_k))$, is $H(P) = -C\sum_{i=1}^{n} p_i \log p_i$, where $C > 0$. This also proves that the uncertainty measurement coincides with $-\log p$. The proof can be seen in Jones (1989: 32).

10. Note that Gibbs' lemma does not specify that number sets a_i and b_i should be probability distributions. Probability distributions are a specific case for the lemma, for which $\sum_{i=1}^{n} a_i = \sum_{i=1}^{n} b_i = 1$ is verified. In Cover and Thomas (2006: 31), we can find a more general inequality, proven based on Jensen's inequality for convex functions. For non-negative numbers (equality if and only if a_i/b_i is constant):

$$\sum_{i=1}^{n} a_i \log \frac{a_i}{b_i} \geq \left(\sum_{i=1}^{n} a_i\right) \log \frac{\sum_{i=1}^{n} a_i}{\sum_{i=1}^{n} b_i}.$$

11. This is a pair because increasing one number by a certain amount means that another must be decreased by the same amount in order to preserve the sum.
12. This is immediately proven by simply taking Gibbs' lemma with $a_i = p_i$ y $b_i = q_i$.
13. When $p_i > 0$ and $q_i = 0$ are verified, we obtain an infinite relative entropy regardless of how small p_i is. We cannot say that this is a true distance because it is not symmetrical: $J(X, Y) \neq J(Y, X)$. This problem can be solved by defining the divergence between X and Y as $D(X, Y) = J(X, Y) + J(Y, X)$. Although it is sometimes called the Kullback–Leibler distance, divergence is still not a true distance because it does not verify the triangular property. To check this, simply take distributions $P_A = (1/4, 1/2, 1/4)$, $P_B = (1/2, 1/4, 1/4)$, and $P_C = (1/3, 1/3, 1/3)$, for which the following is verified: $D(A, B) = 1/2$ bit, $D(A, C) = 1/6$ bit, and $D(C, B) = 1/6$ bit. Therefore, $D(A, C) + D(C, B) = 1/3 < D(A, B) = 1/2$.
14. Cover and Thomas (2006) define mutual information with the expression obtained in Property 1 (Eq. 2.20), and the definition given here appears as a property. The advantage of doing this backwards is that the initial and final states of a hypothetical design process are shown explicitly.
15. The symmetry property (2.25) enables us to write (2.23) as $I(X, Y) \leq H(Y)$.
16. For example, Kullback (1968: 7) refers to the alphabet size algorithm as the Hartley information measure.
17. In the context of Axiomatic Design, 'information content' refers to the uncertainty of the probability of success: $I = -\log p$. If success means finding the only correct solution out of a total of n different solutions, then $I = -\log p = \log n$. In this last case, it can be verified that the information content will coincide with the upper extreme of the entropy of the variable that defines when success is achieved; i.e. the Hartley information measure.
18. For example, an activity can increase (decrease, respectively) the entropy if it increases (decreases, respectively) the number of variables or the size of the alphabets.
19. For example, in the case of a bearing there will be alphabets for the maximum admissible static load, maximum admissible temperature, maximum

admissible rotation regime, minimum admissible reliability, etc. Given a bearing design, it is known (Harris, 2001) that reliability, life and load are related through the Weibull distribution, and are therefore not independent variables (see Chapter 5).

20. This includes all operation and design parameters and all of their possible values. For example, for a bearing there will be an alphabet for each of the following operation and design parameters: rotation regime, axial load, radial load, assembly pressure, lubricant temperature, lubricant viscosity, lubricant mass flow rate, elastic modulus, Young's modulus, superficial hardness of the different materials, number of rolling elements, rolling element diameter, race curvature radius, common radii, radial clearance, etc.
21. For example, they might be inaccessible because the choice of a particular label in a design parameter prevents the choice of another particular label in another design parameter (for example, due to geometric interferences), because the device is unable to generate a response for a certain range of operation labels, or because the operations required to reach that point cannot be executed or completed.
22. The six positions on a gear shift lever in a car, and the two positions on the power saving switch on a TV are examples of these alphabets in the space of definition for the solutions.
23. The pressure of the tires on a vehicle, the output voltage of an operational amplifier for analog signals, and the flight speed of an aircraft are examples of this type of characteristics.
24. An interval of $\mathbb{R}$ is a connected subset. Although it is not necessary, it is convenient to define the labels as intervals because the continuous image of a connected subset is also a connected subset.
25. When the need establishes a functional relationship between two dimensions of $\mathbb{R}^p$ (i.e. the need is characterized by having to verify the functional relationship $l_j = f(l_i)$, where $i \neq j$), it is advisable to consider one of the two variables as an operation parameter. (For example, in the case of a bearing, for a given design and reliability, the bearing lifetime and load supported will be related by a potential law. In this case, the load should be defined as an operation parameter in space X, and the life as a need in W and a response in Y.) This way, the dimension of the space defining the needs is reduced by one dimension to obtain space $\mathbb{R}^{p-1}$, and this dimension will appear in the space of definition for the solution. If this reduction of dimensions is not possible, we must check that l_i and l_j are not two representations of the same characteristic, in which case Axiomatic Design tells us that one of them should be eliminated. If this does not happen either, Axiomatic Design recommends renaming one of them and removing it from the list of needs, to be treated as a constraint.
26. The alphabet {*rejection*, *acceptance*}, which has two elements associated with $(-\infty, \underline{l}_i)\cup(\overline{l}_i, \infty)$ and $[\underline{l}_i, \overline{l}_i]$, respectively, is not considered because one of its elements is not connected. In $\mathbb{R}$, a set is connected if, and only if, it is an interval. As we will see, it is not advisable to select sets that are not connected because they complicate the generation of probability distributions in space Ξ_Z. See Section 2.10.
27. One example of this last case is the rigidity of a precision positioning device: the greater the rigidity, the greater the precision.

28. For the alphabet {*rejection due to lack*, *acceptance*, *rejection due to excess*}, the probability distribution is (0,1,0). Acceptance or non-acceptance by the customer will eventually be established by the probability distribution associated with the customer alphabet in the space of definition for the needs and the relationship between the response and the needs. See Equations (2.36) and (2.39).
29. These are called transfer functions (rather than just functions) because, as we will see, they are responsible for transferring the variability and noise present in the operation and design parameters to the response.
30. The proof is supported by the following results from Calculus. $I \subset \mathbb{R}$ is connected if, and only if, it is an interval. If $f_1, \ldots, f_r$ are the coordinated functions of f, the function f is continuous in C if, and only if, $f_1, \ldots, f_r$ are continuous in C. If $I_i \subset \mathbb{R}$ is a set of intervals, then the set $C = I_1 \times I_2 \times \ldots \times I_q$, obtained by finding the Cartesian product of those intervals, is connected by polygonals. Any set connected by polygonals is connected. If the function g is continuous in C and set C is connected, then the set $g(C)$ is connected.
31. The encoded variables scale the response with the characteristic values set by the customer. One advantage of this variable change is the elimination of physical units; another is the conversion of all values of interest to the customer to numbers of order unity. Consequently, the encoded variables specifying needs and responses can be compared interchangeably. This will enable us, in the following chapters, to define indicators that measure how good a design is.
32. The following condition must be met: $\sum_{e \in \Omega_W} p_e = 1$. There are cases where this condition cannot be verified. For example, the alphabet {*acceptance from the left*, *rejection*, *acceptance from the right*} would respond to distribution (1,0,1), which cannot be a probability distribution. In these cases, the space of definition for needs must be split into two: {*acceptance from the left*, *rejection*, *rejection from the right*} and {*rejection from the left*, *rejection*, *acceptance from the right*} using distributions (1,0,0) and (0,0,1), respectively. This situation might occur in practice when the same list of needs includes motivations from different customers. If the separation were not possible, it would have an unconnected acceptance label that would complicate the calculation of probabilities in this section.
33. Normally, the cases incorporated in the Ξ_X alphabets are all feasible. For this reason, before performing any analysis, the probability distribution can be approached by a uniform distribution: $p_a = 1/|\Omega_X|\ \forall a \in \Omega_X$. In general, however, the definition of the solution as a Cartesian product will allow for combinations of parameters that will lead to unviable executions (whose probability will be zero). There may also be solutions that, while viable, will be labeled as undesirable in a later study, for example, by reducing their probability in space X. See Equation (2.40).
34. This is because both spaces were considered discrete. Although both spaces have a part originating from the discretization of continuous variables, the size of the intervals where the continuous variables operate is lost once the discrete label is associated. Chapter 3 will introduce the concept of differential entropy for continuous variables. We will see that, in general, the entropy of

both spaces changes due to the effect of the transfer functions. (This is the concept of vulnerability introduced by El-Haik and Yang, 1999.) When the variables are not considered discrete, we must take into account the sizes of the intervals when computing entropies. However, the length of an interval can only have meaning when compared to a gauge interval. In the spaces defined, the only interval that can serve as a gauge due to its immutability during the design process is the acceptance interval, $[\underline{l}_i, \overline{l}_i]$, which is in the space of definition for the need. (Not in the space of definition for the solution, where the intervals are created by the designer while creating the space of definition for the solution and establishing the transfer functions.) In order for the gauge interval mentioned here to change, the motivation, needs, or formulation of the needs by the mediator must also change.

35. Normally, without a comparison of the response to the needs, it cannot be known whether a particular response would be accepted. For this reason, before performing any analysis, the probability distribution can be approached using the following distribution: (1/2, 1/2) $\forall a \in \Omega_X$.
36. This probability calculation assumes a uniform probability distribution on the intervals. This hypothesis might seem too drastic; however, it is not problematic because one of the objectives of advanced design techniques is to achieve robust designs. By definition, such designs are insensitive to noise, and therefore to the shape of the distributions.
37. This definition is similar to the one provided by Suh (1990: 156–58): *probability of success=common range/system range.*
38. This definition assumes the statistical independence of the different functional requirements. It also implicitly assumes a uniform probability distribution over the intervals. When the continuous variation of the variables over the intervals is taken into account, both hypotheses are incorrect. El-Haik and Yang (1999) showed that the correlations between the design parameters mean an increase in the differential entropy that is inherited by the functional requirements. In the same way, the structure of the transfer functions modifies the differential entropy. However, the reiterated application of Axiomatic Design throughout the design hierarchy tends to keep the correlations between parameters and the effect of the distribution detail down to minimum values. The expression given in the definition can therefore be used as an initial approach.
39. The sweep of an operation parameter in an article can lead to a subset of articles. In general, the variation (controlled or not) of parameters increases the size of the subsets.
40. Conceptual design would assign uniform distributions to the alphabets. Detailed design would assign a null probability to a series of labels in the alphabet. The article would assign a probability of 1 to one label, and zero to all other labels in the alphabet. The process concludes when all labels with null probabilities are eliminated. However, this does not ensure the success of the article (the probability in (2.39) might be less than 1). Following Equation (2.40), this only ensures that it is the best of the available solutions. See the stop criterion in Section 2.10.
41. If we make use of Axiomatic Design as presented in Chapter 3, we can assign a null probability to those solutions that do not verify the axioms. Because it

is not necessary to reach space Z in order to check these solutions, development, analysis and evaluation costs are reduced. This is one of the sources of value of advanced design theories.

42. Note that space Ω_Y is defined by space Ω_X and the corresponding transfer functions, both fixed by the different solutions provided by the synthesis operator. Although the analysis operator calculates it, its structure is already given by the space for the solutions and the transfer functions.
43. In Chapter 6, the Principle of Minimum generation of Entropy and Information will establish the need to decrease the amount of necessary information as much as possible.
44. Inequality (2.14) establishes that knowing any other Y variable can only reduce the entropy of X. However, it must be emphasized that this only occurs on average. For example, if the value of variable Y is fixed, the conditioned entropy $H(X/Y = y_j)$ can be greater than, less than or equal to the entropy of X. However, the average according to expression (2.14) of these entropies conditioned by a particular value of Y is always less than the entropy of X. Fixing a particular value of a variable can therefore increase the uncertainty of another variable, but when all values of the first are explored, the uncertainty of the second can only be reduced. For example, new evidence could increase uncertainty in a trial, but on average the evidence will reduce uncertainty (Cover and Thomas, 2006: 29). In a design process, a particular erroneous fact could increase uncertainty. On average, however, the complete collection of data will reduce it.
45. The transfer function was only required to be continuous, and therefore does not necessarily have to be derivable. However, the average slopes can always be defined (in absolute values) as the quotient between $\sup(y_i{}^*)_a - \inf(y_i{}^*)_a$ and each of the lengths of intervals $(x_1)_{i_1}, \ldots, (x_{N(X)})_{i_{N(X)}}$, used to generate the Cartesian product identifying article $a \in \Omega_X$. Here, the superscript * reminds us that the interval $(y_i{}^*)_a$ was calculated by modifying one factor at a time. This procedure is explained in the next chapter.
46. This statement can be reinterpreted: The more knowledge the designer has due to his education and experience, the greater the probability that the problem will be correctly defined (Suh, 1990: 7). However, if resource consumption (i.e. efficiency) is also taken into account as a design criterion, some other statements can be questioned: The amount of information acquired in each iteration of the design process should be as high as possible in order to converge to a solution quickly (Suh, 1990: 7).

3

Axiomatic design

Abstract: The two previous chapters discussed the main characteristics and ingredients of the design process. Although these ingredients are difficult to generalize, N. P. Suh (1990) found a common core in what came to be known as Axiomatic Design. His objective, as he himself describes it (Suh, 1990: 4), is to establish a scientific basis for design. The result is a theory capable of generating value because it incorporates, in a natural way, the main needs of any designer: robustness, efficiency, simplicity, and so on, along with a maximum probability of success. The following sections describe the theoretical framework on which this philosophy of design is based. The chapter includes an application of Axiomatic Design during the conceptual design phase of the main bearing configuration of a turbofan.

Key words: independence, information content, coupled and uncoupled design, ideal design, adjustment directions.

3.1 Introduction to axiomatic design

Chapter 2 described the spaces of definition in terms of the general structure of a design process. A flow chart for such a process is shown in Figure 3.1 (compare to Figure 2.3). However, the special way in which Axiomatic Design defines needs in terms of functional requirements and constraints adds a differential element to the space of definition for the needs. This difference is also inherited by the space of definition for the response. In the same way, the design principles established by Axiomatic

Figure 3.1 Flow chart of a design process and spaces of definition

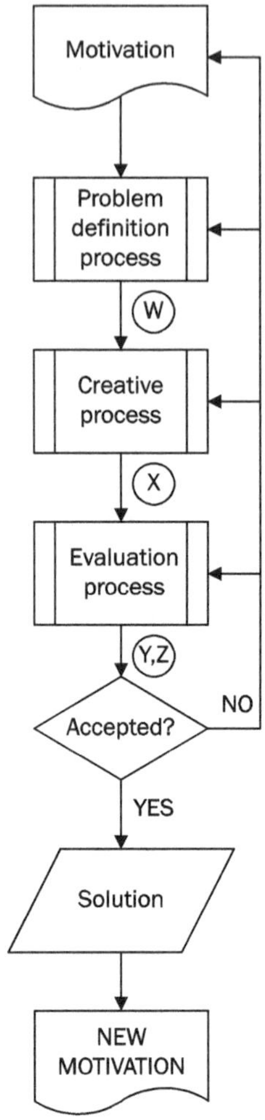

Design provide a special structure for the space of definition for the solution and the transfer functions.

Several examples of Axiomatic Design applications can be found in the literature (Suh, 1990, 2001, 2005; Gebala and Suh, 1992; Hirani and Suh, 2005; Arcidiacono et al., 2006; Conçalves-Coelho and Mourâo, 2007; Thompson, Kwon and Park 2009; Thompson et al., 2009; Zambrano, 2009; Rodríguez-Pastor and Benavides, 2011). Kulak et al. (2010) provide an interesting review of Axiomatic Design applications. The relationship of Axiomatic Design to quality is discussed by Suh (1995),

El-Haik (2005), Yihai et al. (2009) and Tchidi and He (2010), among others. Lee (2006), Lee and Jeziorek (2006) and Cai et al. (2009) have presented strategies for reducing the degree of coupling of the design matrix. Suh (1999, 2005, 2007), Lee (2003), Lu (2009) and Matt (2007, 2009) links Axiomatic Design and complexity. The range of application of the axioms has been increased through the addition of fuzzy operators (Cebi and Kahraman, 2008) and creativity boosting tools (Shirwaiker and Okudan, 2006). Brown (2005), Thompson (2009) and Thompson, Thomas and Hopkins (2009) present very interesting applications of Axiomatic Design to engineering design teaching.

3.1.1 Functional domains and hierarchy of the design process

As we have seen, design is an unstructured problem that must be solved by iterating continuously between 'How shall we do it?' (synthesis) and 'What is achieved?' (analysis). Design is a continuous compromise between what we want to achieve and how we want to achieve it. However, although the questions remain virtually identical in each iteration, the objects to which they refer change constantly. When we analyze the design process from this new perspective, we find a hierarchy that establishes predecessors and descendants for each object. This idea, which is one of the fundamental pillars of Axiomatic Design, assumes that everything in design is, by nature, subject to becoming hierarchical (Suh, 1990: 4). In addition, Axiomatic Design classifies this hierarchy into four domains (Suh, 2001: 10).

1. **Customer domain:** This is characterized by the needs (or attributes) that the customer is seeking for a product, process, system or material.[1] The elements of this domain are called customer needs (CN) or customer attributes (CA).
2. **Functional domain:** This is characterized by the functional requirements (FR) and constraints (CS). It is a reformulation of the needs, or the characterization that the designer makes of the needs perceived by the customer for a product.[2]
3. **Physical domain:** This is characterized by the design parameters (DP). It is a reformulation of the functional requirements in terms of the physical realities that can satisfy them.[3]
4. **Process domain:** This is characterized by the process variables (PV). It is a reformulation of the design parameters in terms of the

processes that can generate the physical realities in the previous domain.[4]

The order of the domains in Axiomatic Design is important because the four domains are related according to this established order: if one domain represents what we want to achieve, the next one represents how to achieve it. The process is also iterative. Once we have information about how to satisfy a particular requirement, we may have to modify what we are seeking. On the other hand, detailing how to do something implies specifying once again what we want to achieve. For this reason, the functional requirements, design parameters and process variables are broken down, resulting in a hierarchy. At the highest point in the hierarchy, the initial functional requirements generate design parameters associated with a solution that must be detailed. In order to detail it, second-level functional requirements are generated, which will in turn require second-level design parameters. The design process continues, layer by layer, until the solution is completely detailed. This process of decomposition and zigzagging among the domains establishes a hierarchy of functional requirements, design parameters and process variables that are a representation of the design architecture. The hierarchy of functional requirements and design parameters in turn produces a hierarchy in the design process (Suh, 1990: 36).

This evidence proves that in the context of Axiomatic Design, functional requirements and design parameters have a hierarchy, and can therefore be broken down. This first postulate of Axiomatic Design also accepts that: 1) the functional requirements on each level cannot be broken down to generate the following level before defining a set of design parameters (i.e. a solution in the physical domain that satisfies the previous level of functional requirements) (Suh, 1990: 36), and 2) an element on a lower level must have a clearly identified predecessor; in other words, a functional requirement may not be introduced on a level of a design hierarchy if it is not supported by a functional requirement on a previous level of the hierarchy (Suh, 2001: 113). Figure 3.2 shows the functional requirements, constraints and design parameters as a function of time in a sequence of several iterations that also enables us to illustrate the design hierarchy. As we will see in Section 3.6, the methodology that accompanies Axiomatic Design also modifies the structure of the design process.

The relationship between the spaces of definition discussed in Chapter 2 and the domains established by Axiomatic Design is shown schematically in Figure 3.3. The spaces of definition for the needs, response and satisfaction

Figure 3.2 **Flow chart of the design process**

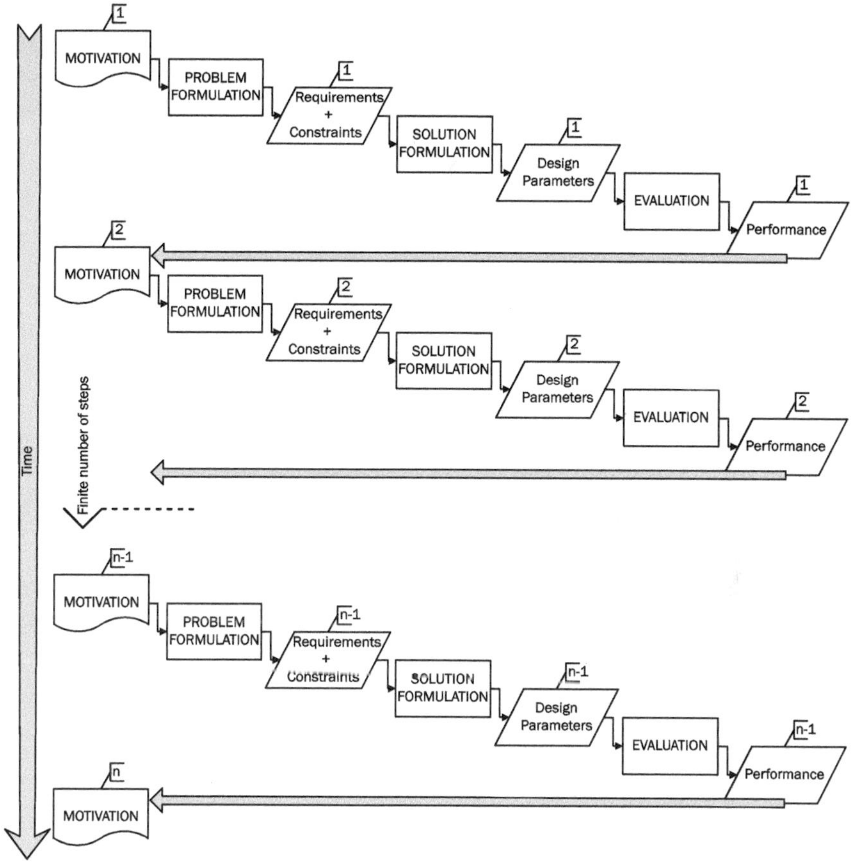

Note: Each stage will produce a level in the design hierarchy in terms of functional requirements, constraints and design parameters.

are within the functional domain; and the physical domain and the process domain are within the space of definition for the solution. As we can see, the mapping between the solution and the response is indispensable in both approaches.[5] As mentioned in Chapter 2, such mapping is controlled by transfer functions. However, Axiomatic Design provides more detail on the structure of the transfer functions by stating that they are a composition of two types of intermediate transfer functions, the ones that link the device to the response $RF = RF(DP)$, and the ones that link the production process to the device $DP = DP(PV)$. As we will see,

Figure 3.3 Spaces of definition and domains in a design process

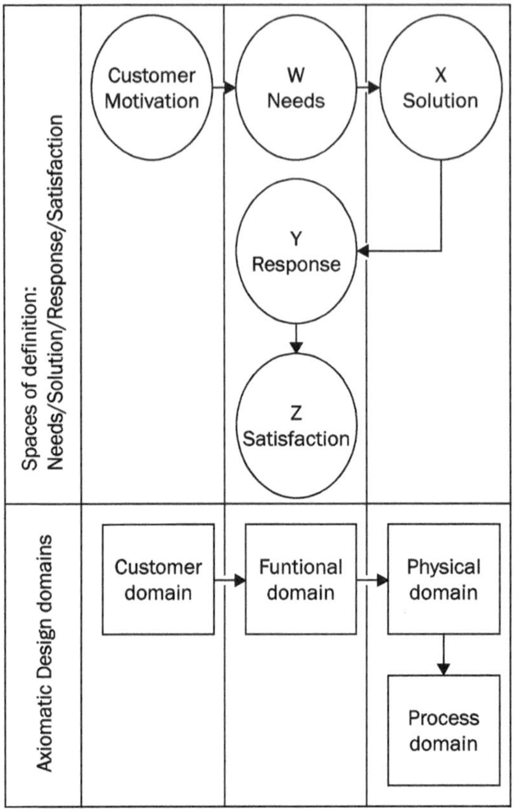

Axiomatic Design applies the same principles to all types of transfer functions.[6]

3.1.2 Transfer functions and design matrix

Transfer functions are the equations for the design. These equations were introduced in Chapter 2 to relate the response margins to the tolerances. They were only required to be continuous functions between spaces of real numbers: $f{:}C \to \mathbb{R}^r$, where $C = [\underline{m}_1, \overline{m}_1] \times [\underline{m}_2, \overline{m}_2] \times \ldots \times [\underline{m}_q, \overline{m}_q] \subset \mathbb{R}^q$ and C is the Cartesian product formed by the intervals where each of the q design parameters vary.[7] Because f is continuous, $f(C)$ is the Cartesian product of r intervals. Each of these intervals is the image of each of the r coordinated functions associated with the r functional requirements. If $f_i(C) \subset \mathbb{R}$ is an interval corresponding to the image of the ith coordinated function in the transfer function, then the image of C is $f(C) = f_1(C) \times f_2(C) \times \ldots \times f_r(C) \subset \mathbb{R}^r$. The probability of success for each functional requirement would be $ip([\underline{l}_i, \overline{l}_i], f_i(C))$, where the function ip is defined as described in (2.36). If all of the requirements were statistically independent, the total probability of success would be the product of the probabilities:[8]

$$p = ip\left([\underline{l}_1, \overline{l}_1], f_1(C)\right) \ldots ip\left([\underline{l}_r, \overline{l}_r], f_r(C)\right) \tag{3.1}$$

If we set all operation parameters except one at point $m_i, \in [\underline{m}_i, \overline{m}_i]$, we can define the subsets

$$C_i = \{m_1\} \times \ldots \times \{m_{i-1}\} \times [\underline{m}_i, \overline{m}_i] \times \{m_{i+1}\} \times \ldots \times \{m_q\} \subset C \tag{3.2}$$

This approach always enables us to write the following matrix:

$$\begin{pmatrix} A_{11} & \cdots & A_{1q} \\ \vdots & \ddots & \vdots \\ A_{r1} & \cdots & A_{rq} \end{pmatrix} = \begin{pmatrix} \dfrac{\sup[f_1(C_1)] - \inf[f_1(C_1)]}{\overline{m}_1 - \underline{m}_1} & \cdots & \dfrac{\sup[f_1(C_q)] - \inf[f_1(C_q)]}{\overline{m}_q - \underline{m}_q} \\ \vdots & \ddots & \vdots \\ \dfrac{\sup[f_r(C_1)] - \inf[f_r(C_1)]}{\overline{m}_1 - \underline{m}_1} & \cdots & \dfrac{\sup[f_r(C_q)] - \inf[f_r(C_q)]}{\overline{m}_q - \underline{m}_q} \end{pmatrix} \tag{3.3}$$

This is a rectangular matrix, size $r \times q$, containing only non-negative elements. As we can see, its elements depend on set C. As long as the transfer function in C is continuous, this matrix always exists. If the transfer function is also differentiable in C, we can construct the Jacobian matrix given by the partial derivatives of the coordinated functions:

$$\begin{pmatrix} A_{11} & \cdots & A_{1q} \\ \vdots & \ddots & \vdots \\ A_{r1} & \cdots & A_{rq} \end{pmatrix} = \begin{pmatrix} \dfrac{\partial f_1(m_1,..,m_q)}{\partial m_1} & \cdots & \dfrac{\partial f_1(m_1,..,m_q)}{\partial m_q} \\ \vdots & \ddots & \vdots \\ \dfrac{\partial f_r(m_1,..,m_q)}{\partial m_1} & \cdots & \dfrac{\partial f_r(m_1,..,m_q)}{\partial m_q} \end{pmatrix} \tag{3.4}$$

This matrix may contain positive, negative or null elements. In a general design process, all elements in a design matrix can have different physical units (such is the case when each functional requirement and each design parameter have different units). When the elements have different dimensions, they are difficult to compare. It is therefore advisable to make use of the encoded variables defined in (2.32) in order to redo the two matrices above without the dimensions from the functional requirements:[9]

$$\begin{pmatrix} A_{11} & \cdots & A_{1q} \\ \vdots & \ddots & \vdots \\ A_{r1} & \cdots & A_{rq} \end{pmatrix} = 2 \begin{pmatrix} \frac{\sup[f_1(C_1)] - \inf[f_1(C_1)]}{(\overline{l}_1 - \underline{l}_1)(\overline{m}_1 - \underline{m}_1)} & \cdots & \frac{\sup[f_1(C_q)] - \inf[f_1(C_q)]}{(\overline{l}_1 - \underline{l}_1)(\overline{m}_q - \underline{m}_q)} \\ \vdots & \ddots & \vdots \\ \frac{\sup[f_r(C_1)] - \inf[f_r(C_1)]}{(\overline{l}_r - \underline{l}_r)(\overline{m}_1 - \underline{m}_1)} & \cdots & \frac{\sup[f_r(C_q)] - \inf[f_r(C_q)]}{(\overline{l}_r - \underline{l}_r)(\overline{m}_q - \underline{m}_q)} \end{pmatrix} \tag{3.5}$$

$$\begin{pmatrix} A_{11} & \cdots & A_{1q} \\ \vdots & \ddots & \vdots \\ A_{r1} & \cdots & A_{rq} \end{pmatrix} = 2 \begin{pmatrix} \frac{\partial f_1(m_1,..,m_q)}{(\overline{l}_1 - \underline{l}_1)\partial m_1} & \cdots & \frac{\partial f_1(m_1,..,m_q)}{(\overline{l}_1 - \underline{l}_1)\partial m_q} \\ \vdots & \ddots & \vdots \\ \frac{\partial f_r(m_1,..,m_q)}{(\overline{l}_r - \underline{l}_r)\partial m_1} & \cdots & \frac{\partial f_r(m_1,..,m_q)}{(\overline{l}_r - \underline{l}_r)\partial m_q} \end{pmatrix} \tag{3.6}$$

The two matrices above show all elements in the same column with the same units, but they could still have columns with different units. One possible way of eliminating this problem is by dividing each column vector by its Euclidean norm.[10] Another possibility is to multiply each column by the half range of variation of the design parameter, in which case we would obtain:

$$\begin{pmatrix} A_{11} & \cdots & A_{1q} \\ \vdots & \ddots & \vdots \\ A_{r1} & \cdots & A_{rq} \end{pmatrix} = \begin{pmatrix} \frac{\sup[f_1(C_1)] - \inf[f_1(C_1)]}{\overline{l}_1 - \underline{l}_1} & \cdots & \frac{\sup[f_1(C_q)] - \inf[f_1(C_q)]}{\overline{l}_1 - \underline{l}_1} \\ \vdots & \ddots & \vdots \\ \frac{\sup[f_r(C_1)] - \inf[f_r(C_1)]}{\overline{l}_r - \underline{l}_r} & \cdots & \frac{\sup[f_r(C_q)] - \inf[f_r(C_q)]}{\overline{l}_r - \underline{l}_r} \end{pmatrix} \tag{3.7}$$

$$\begin{pmatrix} A_{11} & \cdots & A_{1q} \\ \vdots & \ddots & \vdots \\ A_{r1} & \cdots & A_{rq} \end{pmatrix} = \begin{pmatrix} \frac{\overline{m}_1 - \underline{m}_1}{\overline{l}_1 - \underline{l}_1} \frac{\partial f_1(m_1,..,m_q)}{\partial m_1} & \cdots & \frac{\overline{m}_q - \underline{m}_q}{\overline{l}_1 - \underline{l}_1} \frac{\partial f_1(m_1,..,m_q)}{\partial m_q} \\ \vdots & \ddots & \vdots \\ \frac{\overline{m}_1 - \underline{m}_1}{\overline{l}_r - \underline{l}_r} \frac{\partial f_r(m_1,..,m_q)}{\partial m_1} & \cdots & \frac{\overline{m}_q - \underline{m}_q}{\overline{l}_r - \underline{l}_r} \frac{\partial f_r(m_1,..,m_q)}{\partial m_q} \end{pmatrix} \tag{3.8}$$

Therefore, as we can see in the last matrix, the numerical value of the elements in the matrix depends on the permitted variation for the functional requirements, the variation established for the design parameters, and the sensitivity of the transfer functions. By varying any of these ingredients, we adjust the relative importance of one term with respect to the others. If the variation range of a design parameter is reduced, the importance of a column is decreased. If the variation range of a functional requirement is reduced, the importance of a row is increased.[11] By eliminating the non-dominant terms in matrix (3.7) (replacing them with zeros) and marking the dominant terms with an X, we obtain a qualitative version of the design matrix that is widely used in Axiomatic Design. One example could be the following matrix, where $q = 6$ and $r = 5$, in which the non-dominant terms have been replaced by blank spaces.

$$\begin{pmatrix} A_{11} & \cdots & A_{1q} \\ \vdots & \ddots & \vdots \\ A_{r1} & \cdots & A_{rq} \end{pmatrix} = \begin{pmatrix} X & X & X & & X & X \\ & X & & & & \\ & X & & X & & \\ & & X & & & X \\ X & & X & & X & \end{pmatrix} \tag{3.9}$$

This matrix could undergo a reordering procedure by simply permuting the position of the rows and columns. The objective is usually to find a matrix as triangular as possible with a full diagonal.[12]

$$\begin{pmatrix} A_{22} & A_{24} & A_{23} & A_{26} & A_{25} & A_{21} \\ A_{32} & A_{34} & A_{33} & A_{36} & A_{35} & A_{31} \\ A_{12} & A_{14} & A_{13} & A_{16} & A_{15} & A_{11} \\ A_{12} & A_{44} & A_{43} & A_{46} & A_{45} & A_{41} \\ A_{52} & A_{54} & A_{53} & A_{56} & A_{55} & A_{51} \end{pmatrix} = \begin{pmatrix} X & & & & & \\ X & X & & & & \\ X & & X & X & X & X \\ & & X & X & & \\ & & X & & X & X \end{pmatrix} \tag{3.10}$$

In the initial stages of a design process, the matrix is normally written qualitatively, as shown in (3.9). If it has already been reordered, the matrix is written as shown in (3.10). The reason is that during these stages, only the group of experts' qualitative knowledge of the transfer functions is available, for which the precise values are not known. Note that the problem might not be linear, and the operational window might not yet be determined.

3.1.3 Transfer function for small variations

If the transfer function is differentiable, the response can be written in matrix form as $y \simeq y_o + Ax$, where $y \in \mathbb{R}^r$, $y_o \in \mathbb{R}^r$, $x \in \mathbb{R}^q$ and A is the matrix (3.8). The smaller the variation, the better this approach will be.[13]

$$\begin{Bmatrix} y_1 \\ \vdots \\ y_r \end{Bmatrix} \simeq \begin{Bmatrix} (y_o)_1 \\ \vdots \\ (y_o)_r \end{Bmatrix} + \begin{pmatrix} \dfrac{\overline{m}_1 - \underline{m}_1}{\overline{l}_1 - \underline{l}_1} \dfrac{\partial f_1((m_o)_1,..,(m_o)_q)}{\partial m_1} & \cdots & \dfrac{\overline{m}_q - \underline{m}_q}{\overline{l}_1 - \underline{l}_1} \dfrac{\partial f_1((m_o)_1,..,(m_o)_q)}{\partial m_q} \\ \vdots & \ddots & \vdots \\ \dfrac{\overline{m}_1 - \underline{m}_1}{\overline{l}_r - \underline{l}_r} \dfrac{\partial f_r((m_o)_1,..,(m_o)_q)}{\partial m_1} & \cdots & \dfrac{\overline{m}_q - \underline{m}_q}{\overline{l}_r - \underline{l}_r} \dfrac{\partial f_r((m_o)_1,..,(m_o)_q)}{\partial m_q} \end{pmatrix} \begin{Bmatrix} x_1 \\ \vdots \\ x_q \end{Bmatrix} \quad (3.11)$$

The dimensionless components of the vectors are obtained from the corresponding dimensioned values, according to the following expressions:[14]

$$(m_o)_i = \frac{\overline{m}_i + \underline{m}_i}{2}, \quad x_i = \frac{2m_i - \overline{m}_i - \underline{m}_i}{\overline{m}_i - \underline{m}_i} \quad (3.12)$$

$$y_i = \frac{2l_i - \overline{l}_i - \underline{l}_i}{\overline{l}_i - \underline{l}_i}, \quad (y_0)_i = \frac{2f_i((m_o)_1,..,(m_o)_1) - \overline{l}_i - \underline{l}_i}{\overline{l}_i - \underline{l}_i} \quad (3.13)$$

When $y_i \in [-1, +1]$ and $x_i \in [-1, +1]$, the variables fall within the permitted intervals, i.e., $l_i \in [\underline{l}_i, \overline{l}_i]$ and $m_i \in [\underline{m}_i, \overline{m}_i]$ are met. During the design phase, we must choose appropriate values for $(m_o)_i$, which sets the center points for the design parameters (parameter design), and $\overline{m}_i - \underline{m}_i$, which sets the variation range of the design parameters (tolerance design). If the design is not linear, the choice of values $(m_o)_i$ modifies both vector y_o and matrix A. One way to find these values is by maximizing the probability of success.[15]

The extreme values of the response for each functional requirement appear when absolute values are used in the above equation.

$$\begin{Bmatrix} (y_{\pm})_1 \\ \vdots \\ (y_{\pm})_r \end{Bmatrix} \simeq \begin{Bmatrix} (y_o)_1 \\ \vdots \\ (y_o)_r \end{Bmatrix} \pm \tag{3.14}$$

$$\pm \begin{pmatrix} \frac{\overline{m}_1 - \underline{m}_1}{\overline{l}_1 - \underline{l}_1} \left| \frac{\partial f_1((m_o)_1,..,(m_o)_q)}{\partial m_1} \right| & \cdots & \frac{\overline{m}_q - \underline{m}_q}{\overline{l}_1 - \underline{l}_1} \left| \frac{\partial f_1((m_o)_1,..,(m_o)_q)}{\partial m_q} \right| \\ \vdots & \ddots & \vdots \\ \frac{\overline{m}_1 - \underline{m}_1}{\overline{l}_r - \underline{l}_r} \left| \frac{\partial f_r((m_o)_1,..,(m_o)_q)}{\partial m_1} \right| & \cdots & \frac{\overline{m}_q - \underline{m}_q}{\overline{l}_r - \underline{l}_r} \left| \frac{\partial f_r((m_o)_1,..,(m_o)_q)}{\partial m_q} \right| \end{pmatrix} \begin{Bmatrix} |x_1| \\ \vdots \\ |x_q| \end{Bmatrix}$$

For small variations, the above matrix can be replaced by (3.7) to give the following:

$$\begin{Bmatrix} (y_{\pm})_1 \\ \vdots \\ (y_{\pm})_r \end{Bmatrix} \simeq \begin{Bmatrix} (y_o)_1 \\ \vdots \\ (y_o)_r \end{Bmatrix} \tag{3.15}$$

$$\pm \begin{pmatrix} \frac{\sup[f_1(C_1)] - \inf[f_1(C_1)]}{\overline{l}_1 - \underline{l}_1} & \cdots & \frac{\sup[f_1(C_q)] - \inf[f_1(C_q)]}{\overline{l}_1 - \underline{l}_1} \\ \vdots & \ddots & \vdots \\ \frac{\sup[f_r(C_1)] - \inf[f_r(C_1)]}{\overline{l}_r - \underline{l}_r} & \cdots & \frac{\sup[f_r(C_q)] - \inf[f_r(C_q)]}{\overline{l}_r - \underline{l}_r} \end{pmatrix} \begin{Bmatrix} |x_1| \\ \vdots \\ |x_q| \end{Bmatrix}$$

Depending on the sign used, the $(y_{\pm})_i$ components represent the highest or lowest value that can be reached by the response for functional requirement *i*. By undoing the adimensionalization, the following is obtained:

$$\sup(f_i(\bigcup_j C_j)) \simeq f_i\left((m_o)_1,...,(m_o)_q\right) + \frac{1}{2}\sum_{j=1}^{q}\left(\sup[f_i(C_j)] - \inf[f_i(C_j)]\right) \tag{3.16}$$

$$\inf(f_i(\bigcup_j C_j)) \simeq f_i\left((m_o)_1,\ldots,(m_o)_q\right)-\frac{1}{2}\sum_{j=1}^{q}\left(\sup[f_i(C_j)]-\inf[f_i(C_j)]\right) \quad (3.17)$$

$$\sup(f_i(\bigcup_j C_j))-\inf(f_i(\bigcup_j C_j)) \simeq \sum_{j=1}^{q}\left(\sup[f_i(C_j)]-\inf[f_i(C_j)]\right) \quad (3.18)$$

By ensuring that the response falls within the acceptance limits for any design point in set C, the success of the design is ensured. To achieve this, the response intervals must be included within the acceptance intervals. When the length of the acceptance intervals is very small, the following theorem is useful.[16]

Theorem: When the acceptance interval of a functional requirement can be made arbitrarily small, the sum of the columns in matrix (3.7) provides a vector whose components are a lower bound of the inverse of the probability of satisfying each functional requirement.

$$\frac{1}{ip\left([\underline{l}_i,\overline{l}_i],f_i(C)\right)} \geq \sum_{j=1}^{q}\frac{\sup[f_i(C_j)]-\inf[f_i(C_j)]}{\overline{l}_i-\underline{l}_i} \quad (3.19)$$

Proof: The probability of success for functional requirement i is shown in (2.36) and (2.35):

$$ip\left([\underline{l}_i,\overline{l}_i],f_i(C)\right)=\frac{cr\left([\underline{l}_i,\overline{l}_i],f_i(C)\right)}{\sup\left(f_i(C)\right)-\inf\left(f_i(C)\right)} \leq \frac{\min\left(\overline{l}_i-\underline{l}_i,\sup\left(f_i(C)\right)-\inf\left(f_i(C)\right)\right)}{\sup\left(f_i(C)\right)-\inf\left(f_i(C)\right)} \quad (3.20)$$

When the length of the acceptance interval can be made arbitrarily small, we have $\min(\overline{l}_i-\underline{l}_i, \sup(f_i(C))-\inf(f_i(C)))=\overline{l}_i-\underline{l}_i$. On the other hand, because the transfer functions are continuous and sets $\bigcup_j C_j$ and C_j are connected, their images are intervals. Furthermore, because $\bigcup_j C_j \subseteq C$, it must be $f_i(\bigcup_j C_j)\subseteq f_i(C)$ and, consequently, must be verified as follows:

$$\sup\left(f_i(C)\right)-\inf\left(f_i(C)\right) \geq \sup\left(f_i(\bigcup_j C_j)\right)-\inf\left(f_i(\bigcup_j C_j)\right) \quad (3.21)$$

Finally, expression (3.18) enables us to conclude the proof.

$$
\begin{aligned}
ip\left([\underline{l}_i,\overline{l}_i],f_i(C)\right) &= \frac{cr\left([\underline{l}_i,\overline{l}_i],f_i(C)\right)}{\sup\left(f_i(C)\right)-\inf\left(f_i(C)\right)} \\
&\le \frac{\min\left([\underline{l}_i,\overline{l}_i],f_i(C)\right)}{\sup\left(f_i(C)\right)-\inf\left(f_i(C)\right)} = \frac{\overline{l}_i-\underline{l}_i}{\sup\left(f_i(C)\right)-\inf\left(f_i(C)\right)} \\
&\le \frac{\overline{l}_i-\underline{l}_i}{\sup\left(f_i(\bigcup_j C_j)\right)-\inf\left(f_i(\bigcup_j C_j)\right)} \simeq \frac{\overline{l}_i-\underline{l}_i}{\sum_{j=1}^{q}\left(\sup[f_i(C_j)]-\inf[f_i(C_j)]\right)}
\end{aligned}
\tag{3.22}
$$

The sum of the elements in the rows from (3.7) shown in the above expression proves that the solution satisfies the functional requirements within a certain tolerance band of the design parameters (Suh, 1990: 48).[17] The broader the tolerances, or greater the number of design parameters, the lower the upper bound of the probability. A particular case is obtained by taking the probability of success for that functional requirement as equal to 1.0.

$$
1 \ge \sum_{j=1}^{q} \frac{\sup[f_i(C_j)]-\inf[f_i(C_j)]}{\overline{l}_i-\underline{l}_i} \tag{3.23}
$$

On the other hand, if the element of the main diagonal is chosen as the dominant element, the rest of the parameters must meet the following condition:

$$
1 \ge 1-\frac{\sup[f_i(C_i)]-\inf[f_i(C_i)]}{\overline{l}_i-\underline{l}_i} \ge \sum_{\substack{j=1\\ j\ne i}}^{q} \frac{\sup[f_i(C_j)]-\inf[f_i(C_j)]}{\overline{l}_i-\underline{l}_i} \ge 0 \tag{3.24}
$$

When this condition is violated, the probability of satisfying the specifications is lower than 1.0.[18] During the parameter design phase, the variation range of the design parameters is very broad, and the probability of success is low. During the tolerance design phase, the variation range of the design parameters is reduced to increase the probability of success.

When choosing the dominant elements in a design matrix, it is useful to know that:[19] 1) The elements must be compared by rows, not by columns, and 2) Element A_{ij} is dominant when there is no other element A_{ik}, where $k \ne j$, verifying $|A_{ik}| \gg |A_{ij}|$. In general, the dominant elements of a matrix change during parameter design (because the partial derivatives change) and during tolerance design (because the variation range of the parameters changes).

3.1.4 Definition of design

According to N. P. Suh, design involves relating the different domains on each level of the design process hierarchy. In particular, Axiomatic Design defines design as the mapping process from the functional domain to the physical domain, with the aim of satisfying the functional requirements specified by the designer (Suh, 1990: 26). In other words, the objective of the design process is to establish the transfer functions.[20] In particular, Axiomatic Design provides the principles that the mapping technique must apply in order to produce a good design (Suh, 1990: 27).

3.1.5 Functional requirements and constraints

The fundamental underlying idea is that the final design cannot be better than the set of functional requirements it was created to satisfy (Suh, 1990: 29). For this reason, Axiomatic Design presents a very coherent definition of functional requirements. Functional requirements are the characterization that the designer makes of the perceived needs for a product. Suh (1990: 38) defines the functional requirements as the smallest set of independent requirements that completely characterize the design objectives for a specific need.[21]

Chapter 1 established the list of needs as the set of circumstances, conditions, characteristics, qualities and demands that describe the motivation. When these needs are satisfied, the motivation is met. Naturally, the list of needs must be complete, but it does not necessarily have to be a set of independent elements. For the same reason, it does not have to be a minimum set, i.e. a set composed of the smallest number of elements possible.

Definition: A set is said to be comprised of *independent elements* when it is possible to modify or vary any element without the others suffering any kind of modification or variation.

In line with this definition, there are two types of dependencies in the list of needs. The first, which we will call *direct dependency*, occurs when the list of needs is used to establish the alphabets that constitute the space of definition for the needs. The second type, which we will call *indirect dependency*, occurs when the list of needs is used to establish the alphabets in the space of definition for the response.[22] Direct dependencies between the items on the list of needs are present even in the total absence of solutions. On the other hand, indirect dependencies require a solution. In order to determine whether there is a direct dependency, an item on the list of needs is moved, varied or disturbed, and any changes in the other items are studied (in the absence of solutions).[23] To determine whether there is an indirect dependency, an item on the list of needs is moved, varied or disturbed, and the change that would be required in the solution is studied. Finally, the changes that this variation of the solution would cause in the other items on the list of needs are studied.[24]

Definition: The *minimum list of needs* is defined as the set with the smallest number of elements that completely characterizes the list of needs. To reduce the number of items on the initial list, it is necessary to modify, combine, replace and eliminate items from the list.

Definition: The *list of functional requirements* is the broadest subset of independent requirements contained within the minimum list of needs.[25]

Definition: We define *constraints* as those requirements on the minimum list of needs that cannot be added to the set of functional requirements because they would break their independence.

This statement can be deduced from the main characteristic assigned by Suh (1990: 39) to constraints: a constraint does not have to be independent from other constraints and functional requirements. In other words, constraints limit the acceptable design solutions and differ from functional requirements in that they do not necessarily have to be independent (Suh, 2001: 21). At each level in the design hierarchy, the design must satisfy constraints imposed by both external and internal factors (Suh, 2001: 69). There are therefore two types: input constraints, which are constraints on the design specifications, such as cost[26] and physical limits, and system constraints, which are imposed by the system in which the design solution must operate (Suh, 1990: 39). In general, all design decisions at a higher level of the design hierarchy act as (system) constraints for all lower levels (Suh, 2001: 21). In particular, a design parameter on one level of the design hierarchy might act as a constraint on a lower level of the hierarchy. To summarize, system constraints ensure that as one moves down the hierarchy, the design decisions at each level remain consistent with all earlier decisions (Suh, 2001: 31).

Design objectives are described using functional requirements and constraints, while the physical realization is described using design parameters. N. P. Suh's axiomatics do not clearly establish how to generate the list of functional requirements and constraints that serve as a starting point for the design process explained by the theory, but they do provide a good definition of them. Nonetheless, according to

Suh (1990: 30), there are two different approaches: the first for innovation problems, and the second for improvement problems. Because innovating means generating an original design, or creating solutions that did not previously exist, the functional requirements and constraints of an original design must be established in an environment characterized by a complete absence of solutions (solution-neutral environment, Suh, 1990: 30). When defining the functional requirements (and constraints) of an existing design (Suh, 1990: 34), it is advisable to incorporate the customer attributes in a way that determines whether a wrong or incomplete set of functional requirements (and constraints) has been used.

3.1.6 Design principles

Suh (1990: 5) argues that the fact that there are good design solutions and unacceptable design solutions indicates that there are characteristics or attributes that distinguish between good and bad designs. For example, the principle 'the simpler, the better' could be a universally adopted design principle. If it were, it could be taken as an axiom defining the concept 'better.' Axioms are truths that are assumed without proof, and the only way to refute them is by finding a counterexample. As Suh explains (1990: 18), they are formal statements of what people already know, i.e. of the knowledge involved in activities that people engage in routinely. For this reason, if they are correctly stated, it is difficult for anyone to find evidence that they are invalid.

The Axiomatic Design axioms are consistent with the definition of the design problem in terms of functional requirements and constraints. *Axiom 1*: Maintain the independence of the functional requirements. *Axiom 2*: Minimize the information content. Along with the constraints, the axioms limit the set of acceptable solutions. As we will see, in the context of Axiomatic Design, these two axioms enable us to define the ideal design. An *ideal design* has the same number of functional requirements and design parameters, is uncoupled, and has a null information content. This last characteristic is ensured if the system range falls within the design range (Suh, 2001: 116).[27]

If a system was poorly designed, for example, if compliance with the axioms was not ensured, optimization techniques will not always achieve a sufficient degree of improvement. If they do, the operational window will be so narrow that the system will not be robust. In general, a design that does not meet the Independence Axiom cannot be improved upon by optimization as much as a design that does meet the axiom (Suh, 2001: 156). In the following sections, we will study these statements.

3.2 Independence axiom

The functional requirements are a set of independent elements which, along with the constraints, must reflect the customer's motivation as accurately as possible. When a possible solution is provided, an element, which was not available while the problem was being defined, is added. This element is the set of design parameters. Obviously, the solution and the problem are related, so the design parameters and functional requirements are also related. In a general case, the new relationships linking the design parameters and the functional requirements might cause the functional requirements to lose their independence. To avoid this loss of independence between the functional requirements, Axiomatic Design establishes the Independence Axiom, or first axiom.

Independence Axiom: maintain the independence of the functional requirements.

Definition: The solutions that verify the Independence Axiom are described as uncoupled.

Those solutions that break the independence of the functional requirements must be eliminated during the design process in favor of those that maintain the independence. For this reason, this axiom limits the number of solutions to be processed during the design process. To continue with the argumentation introduced in Chapter 2, the designer will mark those solutions that do not meet Axiom 1 with a null probability, thus eliminating the possibility that they will be chosen.

Independence Theorem: The design matrix for a solution that meets the Independence Axiom has no rows without elements, and no column vectors with more than one element.[28]

Proof: If there is a row with no elements, the proposed solution does not resolve the requirement corresponding to that row. If we add a new element to a column in a matrix such as the one described in the theorem, a dependency is created between two functional requirements, violating Axiom 1. By adding more elements, we can arrive at one of two situations: the design matrix either has a maximum rank, or it does not. Section 8 proves that if the rank of the design matrix is maximum, there are new design parameters that make the design matrix the identity matrix (which meets the conditions in the theorem). If it has no maximum rank, at least one functional requirement is not resolved by the solution. The theorem is therefore true.

By reordering the matrix defined by the above theorem by rows and columns, we can always place the columns whose non-null elements do not affect the first functional requirement first, then the columns affecting the second functional requirement, and so on. It is advisable to define sets D_i, containing all design parameters affecting functional requirement i, as follows:

$$D_i = \bigcup_{j=1}^{q} \delta_i(C_j) \subseteq C \tag{3.25}$$

$$\delta_i(C_j) = \begin{cases} \varnothing \ \ if \ \sup(f_i(C_j)) = \inf(f_i(C_j)) \\ C_i \ \ if \ \sup(f_i(C_j)) \neq \inf(f_i(C_j)) \end{cases} \tag{3.26}$$

The imposition of the Independence Axiom and sets (3.25) make the design matrix (3.7) square and diagonal, as shown in (3.27).

$$\begin{pmatrix} A_{11} & \cdots & A_{1r} \\ \vdots & \ddots & \vdots \\ A_{r1} & \cdots & A_{rr} \end{pmatrix} = \begin{pmatrix} \dfrac{\sup[f_1(D_1)] - \inf[f_1(D_1)]}{\overline{l}_1 - \underline{l}_1} & & 0 \\ & \ddots & \\ 0 & & \dfrac{\sup[f_r(D_r)] - \inf[f_r(D_r)]}{\overline{l}_r - \underline{l}_r} \end{pmatrix} \tag{3.27}$$

The design matrix given in (3.8) is the Jacobian matrix of a certain function and, as such, is a 2nd-order tensor. Because tensors can be diagonalized through a coordinate transformation, the design matrix could be made diagonal. Section 3.8 explains the conditions that must be met by the matrix, and how this must be done. However, this coordinate transformation might not be useful in designs with a certain complexity because the new coordinates would simply be a juxtaposition of elements without meaning or physical significance (Suh, 2001: 113). For example, functional requirements are mathematically subject to a possible coordinate change, but their definition as an independent set advises against a coordinate transformation. Similarly, a coordinate transformation of the design parameters would cause the original design parameters to depend on the final ones, complicating the physical realization of the device, or resulting in parameters that do not represent a physically feasible reality (Suh, 2001: 20).[29] However, in other cases, coordinate transformations enlighten about how to solve the design problem without violating the axioms. One example in which this diagonalization process can be carried out by changing the design parameters is the flow and temperature control explained in Section 3.9. However, the best strategy for avoiding the complexity that can result from transforming the design parameters through diagonalization is to seek a diagonal matrix from the same conception as the solution, as described in Section 3.7 for selecting the bearing configuration for a turbofan or in Chapter 6 for obtaining a fuel metering system.

3.3 Information axiom

The Information Axiom establishes that the best design, of those that satisfy the Independence Axiom, is the one requiring the lowest information content to satisfy the functional requirements. Therefore, the design parameters leading to the lowest information content are the best (Suh, 2001: 73).

Information Axiom: Minimize the information content.

Definition: As defined by Nam P. Suh (2001: 71, and 1990: 65) the *information content* is the uncertainty associated with the probability of success. (See Eq. (2.1).)

This definition makes it possible to establish that the design with the highest probability of satisfying the functional requirements is the best design (Suh, 2001: 68).

Definition: If $pdf(FR_1,...,FR_r)$ is the joint probability distribution associated with the functional requirements, and V is the volume that meets the acceptance conditions, then the *probability of success* is:[30]

$$p = \frac{\int_V pdf(FR_1,\ldots,FR_r)\,dFR_1\ldots dFR_r}{\int_{\mathbb{R}^r} pdf(FR_1,\ldots,FR_r)\,dFR_1\ldots dFR_r} = \int_V pdf(FR_1,\ldots,FR_r)\,dFR_1\ldots dFR_r \quad (3.28)$$

To satisfy the Information Axiom, the information content of the design must be minimized. To achieve this, according to definition (2.1), the probability of success (3.28) must be at its maximum. An extreme case is one in which the actual variation of the system's functional requirements (system range) falls within the range specified as acceptable (design range). The necessary and sufficient condition for satisfying the Information Axiom is that the system range must fall completely within the specified design range (Suh, 2001: 69). Furthermore, when the system range is completely included in the design range, the probability of meeting the functional requirements is equal to 1.0, independently of the probability density function (Suh, 2001: 73). For this reason, uniform density functions can be used in the probability calculation, as mentioned in the discussion of expression (2.36). On the other hand, if Axiom 1 has been verified, all of the functional requirements maintain their independence. In other words, the design parameter affecting one requirement does not affect any other functional requirement. Thus, in an uncoupled design, the functional requirements are kept independent of each other: the information content of an uncoupled design is the sum of the information content of each functional requirement (Suh, 2001: 146). This same condition establishes that in an uncoupled design, the joint probability of success is the product of the probability of success of each functional requirement. Consequently, the axioms enable us to establish the probability of success calculation as shown in expression (3.1). The information content comes from replacing (3.1) and (2.36) in (2.1).[31]

$$I = -\log p = -\sum_{i=1}^{r} \log ip\left([\underline{l}_i, \overline{l}_i], f_i(C)\right) = \sum_{i=1}^{r} \log \frac{\sup\left(f_i(C)\right) - \inf\left(f_i(C)\right)}{cr\left([\underline{l}_i, \overline{l}_i], f_i(C)\right)} \quad (3.29)$$

The Information Axiom establishes that the best design that meets the Independence Axiom is the one that minimizes expression (3.29). This is achieved when the variation of response $[\inf(f_i(C)), \sup(f_i(C))] = f_i(C)$ is completely included in acceptance interval $[\underline{l}_i, \overline{l}_i]$ for each functional requirement i. Note that the argument of the logarithm in (3.29) is bound by expression (3.19), which was obtained assuming that each acceptance interval be arbitrarily small.

$$I \geq \sum_{i=1}^{r} \log \sum_{j=1}^{q} \frac{\sup[f_i(C_j)] - \inf[f_i(C_j)]}{\overline{l}_i - \underline{l}_i} \quad (3.30)$$

Note that the arguments in the second sum in (3.30) are the elements from design matrix (3.7). Remember that this bound requires that the acceptance intervals be arbitrarily small. If the acceptance intervals were arbitrarily large, we would have $I = 0$. On the other hand, if the design verifies the Independence Axiom, relationships (3.2), (3.25) and (3.26) enable us to write the information content as:

$$I = \sum_{i=1}^{r} \log \frac{\sup\left(f_i(D_i)\right) - \inf\left(f_i(D_i)\right)}{\overline{l}_i - \underline{l}_i} \quad (3.31)$$

From this point on, the only way to reduce the information content is by decreasing the length of the response interval. Note that, as established by (3.25) and (3.31), the addition of new design parameters to the whole can never reduce the length of the response interval, but it can increase it.[32]

Information Theorem: The design matrix for a solution that meets the Independence Axiom and the Information Axiom for an arbitrary set of acceptance intervals has no row vector containing more than one element, and no column vector containing more than one element.

Proof: Simply use the Independence Theorem and expression (3.31) along with (3.25) and (3.21).

The easier way to meet the Information Axiom is by seeking the extreme case, in which all of the operation and design parameters are perfectly known. Given the impossibility of meeting this objective in a real design environment, the designer sets strict tolerances for the operation and design parameters, and prohibits or shields those sources of noise in the environment that exceed certain intensities. This design strategy is costly and not very robust. First the Independence Axiom, and later the Information Axiom, make it possible to accommodate higher tolerances and noise sources in the design parameters. The Information Axiom provides the theoretical basis for the optimization and robustness of the design (Suh, 2001: 39).

3.4 Independence of the axioms

The independence of the two axioms is an interesting issue, not to mention a fundamental question. Suh (2001: 114) warns of two false arguments for reducing the set of axioms. Because an uncoupled design has fewer addends in expression (3.30) than a coupled design, we could conclude: 1) 'An uncoupled design has a lower information content than a coupled design', and 2) 'One way of reducing the information content is by decoupling the design.' However, both statements are false. Statement 1) is not true because there are other ways of reducing the information other than independence. For example, a design window can be found in a coupled design which, due to the special values of its derivatives, leads to a lower information content. Statement 2) is false because the information content of an uncoupled design can be increased by simply setting the variation of the design parameters and increasing the amplitude of the elements in the design matrix. (For convenience, think of a design with a single functional requirement and a single design parameter related according to $FR = kDP$. In this case, all we have to do is systematically increase k for the variation of FR to grow indefinitely.)

This degree of freedom is not considered in the Information Axiom. To correctly prove the independence of the axioms, we must resort to the concept of differential entropy of a continuous variable.

3.4.1 Entropy of a continuous variable

Let $x_i \in \mathbb{R}$, where $i = 1,2,\ldots,n$, be a set of real variables. Let $pdf(x_1,\ldots,x_n)$ be the joint probability density function for all of them. The probability that variable $x = (x_1,\ldots,x_n) \in \mathbb{R}^n$ is within volume $V \subset \mathbb{R}^n$ defined by intervals $(x_i, x_i + dx_i)$ is

$$P(V) = \int_V pdf(x_1,\ldots,x_n)\,dx_1\ldots dx_n = \int_V pdf(x)\,dV \qquad (3.32)$$

Obviously, the probability that variable x is at some point in its definition field must be one; in other words:

$$P(\mathbb{R}^n) = \int_{\mathbb{R}^n} pdf(x_1,\ldots,x_n)\,dx_1\ldots dx_n = \int_{\mathbb{R}^n} pdf(x)\,dV = 1 \qquad (3.33)$$

If we take the partition

$$\wp = \left\{ V_j \subset \mathbb{R}^n : j = 1,\ldots,m \mid V_j \cap V_{k\neq j} = \varnothing \wedge \bigcup_{j=1}^{m} V_j = \mathbb{R}^n \right\}$$

the continuous variable x defines a new discrete variable, X, which takes it values from the labels V_j in alphabet $\wp$. This random variable can be assigned the probability distribution $(p_1,\ldots,p_m)$, where each value in the distribution is given by $p_i = P(V_i)$. The entropy of variable X is therefore:

$$H(X) = -\sum_{j=1}^{m} p_j \log p_j = -\sum_{j=1}^{m} P(V_j)\log P(V_j) \qquad (3.34)$$

If the volume of each element is made to be dV, the average theorem enables us to write each of the probabilities in the above expression as follows (to do so, the density function must be continuous):

$$P(V) = pdf(x \in V)dV \qquad (3.35)$$

The entropy would be:

$$H(X) = -\sum_{j=1}^{m} pdf(x \in V_j)dV \log\left[pdf(x \in V_j)dV\right] \qquad (3.36)$$

If the number of elements in the partition is now made to tend to infinity, and the volume of each element is made infinitesimally small, this expression would be the formal definition of an integral if the logarithmic term did not diverge when $dV \to 0$. By making $m \to \infty$ and $dV \to 0$, the discrete variable X tends to the continuous variable x. For this reason, the entropy of a continuous variable cannot be completely analogous to the entropy of a discrete variable. The following definition is adopted as the entropy of a continuous variable:

$$h(x) = -\int_{\mathbb{R}^n} pdf(x)\log\left[pdf(x)\right]dV \tag{3.37}$$

As we can see, the logarithm does not include the probability, but rather the probability density. For this reason, some books (Cover and Thomas, 2006: 246) do not refer to this expression as entropy, but differential entropy. This means that a continuous variable can have a negative differential entropy (not satisfied with the entropy of a discrete variable). Thus, given a uniform density function in the interval (a, b), the following must be verified:

$$h(x) = -\int_{\mathbb{R}} pdf(x)\log\left[pdf(x)\right]dV = -\int_a^b \frac{1}{b-a}\log\frac{1}{b-a}dx = \log(b-a) \tag{3.38}$$

This can be greater than, less than, or equal to zero if $b - a$ is respectively greater than, less than, or equal to one. The greater the interval, the greater the differential entropy, and the lower the interval, the lower the differential entropy. Because the standard deviation of the uniform distribution is $(b-a)/(2\sqrt{3})$ and the logarithm is a monotonically increasing function, the differential entropy and the standard deviation are related. On the limit $b \to a$, the probability distribution tends to the distribution given by the Dirac Delta, and the differential entropy is $-\infty$. This result indicates that the average uncertainty of a variable whose probability distribution is a Dirac Delta distribution is $-\infty$, and that of a variable uniformly distributed along the entire real line is $+\infty$. For a normal distribution centered on a and with standard deviation s according to (3.39), the differential entropy is given by (3.40).

$$pdf(x) = \frac{1}{s\sqrt{2\pi}}e^{-\frac{1}{2}\left(\frac{x-a}{s}\right)^2} \tag{3.39}$$

$$h(x) = -\int_{\mathbb{R}} pdf(x)\ln\left[pdf(x)\right]dV = \ln\left(s\sqrt{2\pi e}\right) \tag{3.40}$$

In other words, the average uncertainty grows as the logarithm of the standard deviation of the distribution. Consequently, it can be said that the greater the average uncertainty, the greater the interval width and standard deviation (and vice-versa).[33]

When the variable $x = (x_1, \ldots, x_n) \in \mathbb{R}^n$ is transformed into another variable $y = (y_1, \ldots, y_n) \in \mathbb{R}^n$ using the function $y = g(x)$, which meets the conditions of the inverse function theorem so that $x = g^{-1}(y)$ is also verified, the probability density function $pdf_x(x)$ induces a probability density function in the variable y, given by:

$$pdf_y(y) = pdf_x(g^{-1}(y))|J(g^{-1})| \tag{3.41}$$

In this expression, $J(g^{-1})$ is the determinant of the Jacobian matrix obtained by making its elements equal to $\partial g^{-1}_i(y)/\partial y_j$. The function g can be understood as the transfer function that generates response y for a particular value of operation and design parameters x. If $J(g^{-1})$ is a constant value that does not depend on the coordinates, we have the following:

$$\begin{aligned} h(y) &= -\int_{\mathbb{R}} pdf_y(y)\log\left[pdf_y(y)\right]dV_y \\ &= -\int_{\mathbb{R}} pdf_x(g^{-1}(y))\left|J(g^{-1})\right|\log\left[pdf_x(g^{-1}(y))\left|J(g^{-1})\right|\right]dV_y \end{aligned} \tag{3.42}$$

$$h(y) = h(x) - \log|J(g^{-1})| \tag{3.43}$$

Where we have taken into account that:

$$dV_y = \frac{dV_x}{\left|J(g^{-1})\right|} \tag{3.44}$$

Therefore, a transformation that expands the volumes tends to increase the differential entropy, while one that contracts the volumes tends to reduce it. Consequently, transformations change the standard deviations of the distributions associated with the functional requirements, and therefore the probability of success. This argument, used by El-Haik and Yang (1999), enables us to state that Axiom 1 is not a consequence of Axiom 2 because the information content of an uncoupled design can be reduced as desired.[34] Axiom 2 is not a consequence of Axiom 1 because the information content of an uncoupled design can be increased as desired.

In the absence of Axiom 1, a coupled design could be chosen that contains less information than an uncoupled design (Suh, 1990: 67). In the absence of Axiom 2, it would always be possible to conceive an uncoupled design with a greater information content than another uncoupled design. This implies that both axioms are necessary because each one addresses a different conceptual weakness (Suh, 2001: 177).

3.5 Most relevant theorems and corollaries

This section discusses some of the most interesting corollaries and theorems presented by Suh (1990 and 2001).[35] Please note that these corollaries and theorems must be satisfied as long as their use does not violate the functional requirements and constraints.

Theorem 1: The number of design parameters may not be lower than the number of functional requirements.

Proof: In this situation, if the parameters are capable of controlling all of the functional requirements, the design will be coupled, and will violate Axiom 1. If they are not capable of controlling all of the requirements, the design will not meet the specifications.

Corollary 1: In each column of the design matrix, reduce the number of dominant elements.[36]

Proof: For a variation of the design parameters, arbitrary in direction, the greater the number of elements in each column vector, the lower the degree of compliance with Axiom 1.

Corollary 2: For an arbitrary acceptance interval length, reduce the number of functional requirements.

Proof: The information content can be expressed based on expression (3.1) as a sum of positive or null terms. The greater the number of requirements, the greater the number of addends in (3.29). When the number of positive addends is reduced, the information content decreases (as required by the Information Axiom). If all of the addends are null, nothing can be concluded because the information content does not change when functional requirements are added or removed. However, if the length of the acceptance intervals is arbitrary, it can be made as small as desired. The length of the common range can therefore be reduced arbitrarily. This would turn the null terms in expression (3.1) into positive terms (see Eq. 2.36). For this reason, they must be removed.

Corollary 2(*): For an arbitrary acceptance interval length, reduce the number of design parameters.

Proof: Let $y = \sum_{i=1}^{q} A_i x_i$ be a functional requirement that depends linearly on q design parameters x_i, which are statistically independent. Under these conditions, the variance of the functional requirement is $V(y) = \sum_{i=1}^{q} A_i^2 V(x_i)$. Therefore, the standard deviation of the functional requirement increases as the number of design parameters increases, which also increases the range of variation of the system. This causes the probability of success to decrease. According to this expression, Axiom 2 requires reducing the number of design parameters and the variance of the design parameters. (The arbitrary acceptance interval length makes it necessary to reduce the number of parameters with a non-null variance that affect the requirement.) Note that the values of A_i were kept fixed in the proof. If there is an acceptable design with more design parameters and greater parameter variance, but with lower constant values, the appropriate design may be the one with the highest number of parameters. This corollary applies to the elimination of independent design parameters; therefore, the elimination or insertion of a parameter does not change the statistical distribution or the constants of the other parameters in the model. If there is more than one functional requirement, but Axiom 1 is satisfied, the proof performed with a single requirement is applicable.

Theorem 4: The ideal design has the same number of design parameters and functional requirements.

Proof: Theorem 1 states that the number of design parameters must be greater than or equal to the number of functional requirements. Corollary 2(*) states that the number of design parameters must be minimal. Both numbers must therefore coincide.

Corollary 3: Reduce the number of parts or elements.

Proof: As long as reducing the number of parts or elements does not result in non-compliance with Axiom 1, the smaller the number of parts, the smaller the number of design parameters necessary to define them. Corollary 2(*) concludes the proof.

Corollary 4: Use standardization.

Proof: This is obtained from Corollary 2(*), as standardization reduces the number of design parameters and their variability.

Corollary 5: Use symmetries. Symmetrical parts have fewer design parameters than non-symmetrical parts.

Corollary 6: Use the broadest acceptance limits possible.

Proof: The higher the acceptance limits, the greater the common range and the higher the probability of success.

Corollary 7: Find an uncoupled design with a null information content.[37]

Proof: Theorem 4 specifies that the number of design parameters and functional requirements must coincide. Corollary 1 specifies that the columns must contain a single element. Corollary 2(*) specifies that the rows must contain a single element. The matrix can therefore be diagonalized by reordering rows and columns. Corollary 2 specifies that the number of elements on the diagonal must be minimal. Axiom 2 specifies that the information content must be minimal. This minimum is obtained when the system range is within the design range. The ideal design therefore has a diagonal design matrix and a null information content.

Theorem 19: Uncoupled designs are more robust than coupled designs.

Proof: A coupled design has at least one column in the design matrix with two dominant elements. This implies that there is one row with at least two dominant elements, thus violating Corollary 2(*). This means that there is at least one functional requirement with greater variance than if the design were not coupled. Consequently, the capacity to accommodate unforeseen sources of noise is reduced in the coupled design. Robustness is a consequence of the Information Axiom.

Theorem 9: When the design axioms are satisfied for the product and the manufacturing process, the robustness of the production is less compromised (Suh, 1995; Suh, 1990: 41; Suh, 2001: 61).

Proof: Simply apply Theorem 19 to the product design and process design.

3.6 Design process

Figure 3.4 shows a flow chart for a design process incorporating the Axiomatic Design decision criteria. As we can see, it differs from a classic flow chart (Figure 3.1) due to the addition of four assessment and evaluation stages with decision criteria based on the Axioms, Theorems and Corollaries discussed in this chapter. Nevertheless, it is worth emphasizing that through the design process, these axioms, corollaries and theorems must always be kept in mind and applied whenever possible (Suh, 1990: 64).

The essential elements of an Axiomatic Design process shown below have been reformulated based on those introduced by Suh (2001: 68):

1. There are four domains: the customer domain, functional domain, physical domain, and process (manufacturing or production) domain.
2. Proper statement of the customer's needs in the customer domain.
3. Mapping of the customer's needs from the customer domain to the functional domain in a solution-neutral environment, in order to determine the functional requirements and constraints. At this point, Corollary 2 is useful.

Figure 3.4 **Flow chart for a design process including the axiomatic design decision criteria**

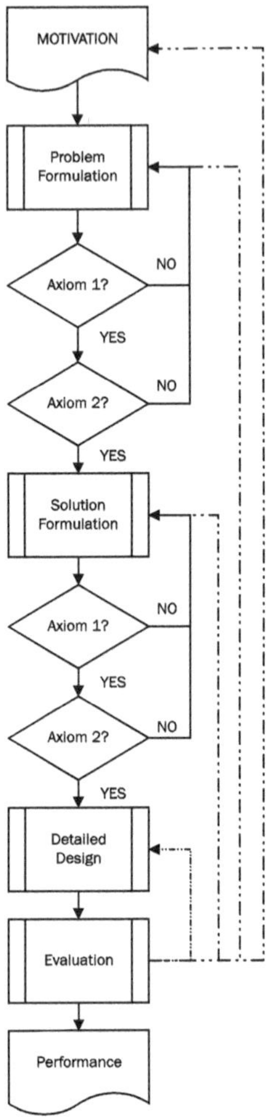

4. Mapping from the 'what' in the functional domain to the 'how' in the physical domain to establish the design parameters that satisfy the functional requirements for product design, and the process variables that satisfy the design parameters for process design. At this point, Theorem 4 and Corollary 1 are useful for finding a design with the best design matrix. For the process of seeking creative solutions, the most important terms are those outside the diagonal (Suh, 1990: 62) because seeking their elimination promotes new ways of thinking.

5. The need to satisfy the axioms during the previous mappings. A real design process always starts with Axiom 1 and seeks an uncoupled design. For choosing between several designs that satisfy Axiom 1, Axiom 2 is used to determine which of the proposed designs is the best (Suh, 1990: 66).
6. Iterating (zigzagging) between the domains in order to break down the functional requirements, design parameters and process variables. Establishing the design hierarchy through the decomposition process until the design is complete.
7. If the resulting information content is not null, optimizing the design using the Information Axiom.[38]

3.7 Example application in the aeronautical industry: main bearing configuration on a jet engine

This example uses Axiomatic Design methodology for solving the following conceptual design problem (adapted from Benavides, 2011).[39]

> **Motivation:** To establish the optimum configuration for the main bearings that will support the different shafts in a turbojet or turbofan engine.

The engine shown in Figure 3.5 illustrates the type of problem to be solved. This engine has three shafts: the central shaft supports the fan and low-pressure turbine, and the other two shafts are equipped with a compressor in front and a turbine in back. Normally, the three shafts would be axially pushed due to the intake of gas in the engine. On the other hand, the engine housing will be retained by the vehicle's aerodynamic resistance, and subjected to transverse forces during rotational maneuvers. Figure 3.6 shows the main forces acting on the three shafts and housing. Logically, unless the corresponding mechanical elements are introduced between the shafts and housing, their relative position will change due to the effects described above. A need therefore arises.

> **Need:** To maintain the position of the different shafts (relative to the housing) without introducing any torque in the direction of the rotation axis.

Figure 3.5 Drawing of the Trent 500 engine used on the Airbus A340–500 and –600. Courtesy of Rolls-Royce

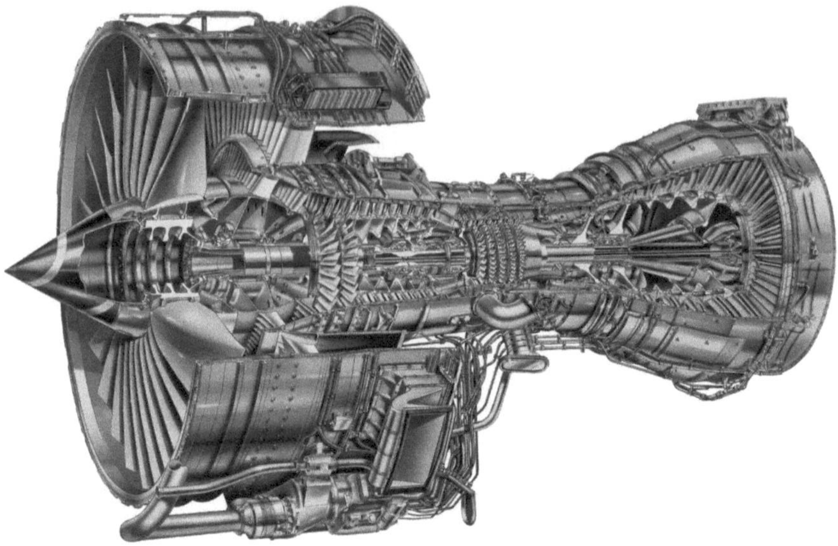

Figure 3.6 Diagram of a three-shaft turbojet showing the mechanical loads that appear on the three coaxial shafts and housing in the absence of supports

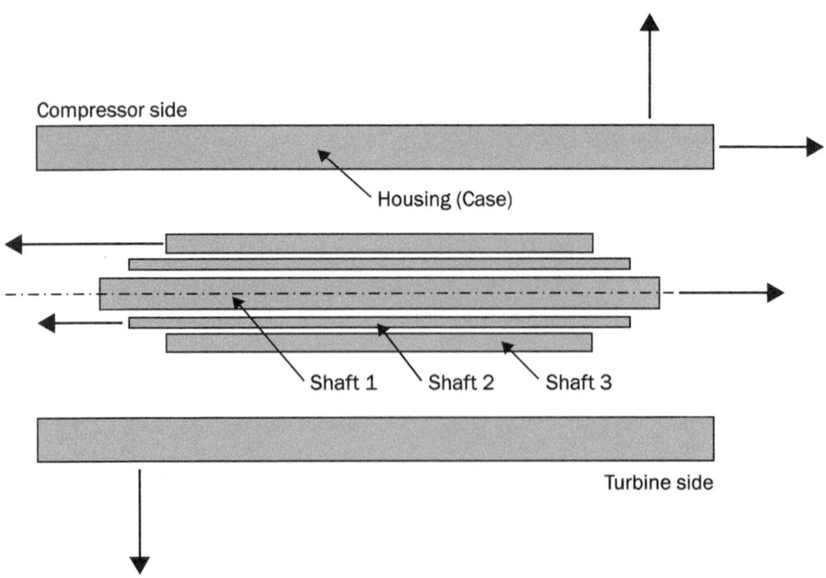

3.7.1 Formulation of the problem. Functional requirements and constraints

Considering the shafts as solid parts, rigid body theory warns us that each shaft has six degrees of freedom. For this reason, the motivation can be written using the following list of needs.

List of needs for each shaft (list 1): Prevent axial translation, prevent radial translation, prevent radial rotation, and allow axial rotation.

In order for this list of needs to be a set of functional requirements, it must be minimal and independent. As we can see, list 1 is not independent. Indeed, the introduction of radial rotation at an arbitrary point will change the radial position of certain points on the part. To solve this, there are two possibilities.

List of needs for each shaft (list 2): Prevent axial translation, prevent radial translation of point A on the shaft, prevent radial rotation of the shaft around point A, and allow axial rotation.

List of needs for each shaft (list 3): Prevent axial translation, prevent radial translation of point A on the shaft, prevent radial translation of point B on the shaft, and allow axial rotation.

The second list is still not independent because the translation and rotation depend on each other through the chosen point A. This is solved in list 3 because the translations at points A and B are independent for small displacements (one of the points can be displaced transversely without displacing the other). However, if the shaft has a finite diameter, points A and B must be on the surface of the shaft, in which case the axial rotation is coupled with the radial displacement of both points (a 90° rotation with respect to the axial shaft changes the radial position of points A and B). The need to 'allow axial rotation' is therefore coupled with the needs to 'prevent radial translation of A' and 'prevent radial translation of B.' This must therefore be considered a constraint. The problem would be expressed as follows:

List of functional requirements: Prevent axial translation, prevent radial translation of a point on the shaft (point A), and prevent radial translation of another point on the shaft (point B).

List of constraints: Allow axial rotation.

This list of functional requirements can be repeated for each of the shafts while remaining independent.[40]

3.7.2 Formulation of a solution. Design parameters

A mechanical solution that meets the constraint and verifies the functional requirements is the introduction of a rolling element between the shafts and/or housing that prevents axial and radial slip. These rolling elements are the roller bearings. In general, bearings introduce axial force, radial force and radial torque at the point on the shaft where they are mounted. We can therefore deduce that each bearing introduces three design parameters: axial force, radial force and radial torque. However, because their thickness is much less than the length of the shaft they must hold, they allow only small radial torques. Table 3.1 shows a transfer function expressed as a design matrix, where capital Xs represent dominant effects and lower-case Xs represent effects of a lower order of magnitude than the dominant ones. The design matrix was constructed with arbitrary bearing positions (that do not necessarily coincide with points A and B), so the radial force of a bearing affects the radial position of both points.

Table 3.1 Design matrix relating the design parameters for an arbitrary set of bearings to the functional requirements of the shaft where they are mounted

		Design parameters									
		Bearing 1			Bearing 2			Bearing 3			. . .
		Axial force	Radial force	Radial torque	Axial force	Radial force	Radial torque	Axial force	Radial force	Radial torque	. . .
Functional requirements for shaft 1	Axial displacement	X			X			X			. . .
	Radial displacement of point A		X	x		X	x		X	x	. . .
	Radial displacement of point B		X	x		X	x		X	x	. . .

3.7.3 Application of the Independence Axiom

Note that the number of bearings on the shaft and their position are values that the designer must determine using some kind of decision-making procedure. According to Axiomatic Design, this is not an optimum design because it has more design parameters than functional requirements, and because the design matrix is coupled. To improve the design, all of the Xs not located on the main diagonal must be eliminated. To do this, we will proceed row by row. On each row, we will keep the first dominant X that enables us to construct a main diagonal on the matrix. The result is shown in Table 3.2.

The design matrix obtained by applying the axioms indicates that: 1) only two bearings must be used, 2) the radial force introduced by a bearing must only affect the radial displacement of one of the points, and 3) only one of them must support axial forces. Point 1) requires the use of a pair of bearings as the design solution. Point 2) can be achieved if bearing 1 is displaced to point A, and bearing 2 to point B. Point 3) can be achieved if bearing 1 is a ball bearing and bearing 2 is a cylinder bearing. This solution is therefore feasible.

If there is an arbitrary number of shafts, this solution indicates that a pair of bearings as described above is needed for each shaft. However, it says nothing about whether it is better to fix the bearings to the housing

Table 3.2 Design matrix that satisfies the axioms

		Design parameters									
		Bearing 1			Bearing 2			Bearing 3			...
		Axial force	Radial force	Radial torque	Axial force	Radial force	Radial torque	Axial force	Radial force	Radial torque	...
Functional requirements of shaft 1	Axial displacement	X									...
	Radial displacement of point A		X								...
	Radial displacement of point B					X					...

Table 3.3 **Design matrix for the configuration shown in Figure 3.7**

		Design parameters								
		Pair 1			Pair 2			Pair 3		
		Axial force	Radial force	Radial force	Axial force	Radial force	Radial force	Axial force	Radial force	Radial force
Functional requirements of shaft 1	Axial displacement	X			X					
	Radial displacement of point A		X			X				
	Radial displacement of point B			X			X			
Functional requirements of shaft 2	Axial displacement				X			X		
	Radial displacement of point A					X			X	
	Radial displacement of point B						X			X
Functional requirements of shaft 3	Axial displacement							X		
	Transverse displacement of point A								X	
	Transverse displacement of point B									X

or to another shaft. For example, Table 3.3 shows the design matrix for the configuration shown in Figure 3.7.

Again, Axiomatic Design finds this solution unacceptable because pair 2 simultaneously affects the functional requirements for shafts 1 and 2 (the same is true for pair 3 and shafts 2 and 3). The procedure for diagonalizing the matrix is the same as before: by maintaining the dominant effect that turns the matrix into a diagonal. The result is:

Figure 3.7 Diagram showing a design solution in which each pair of bearings is anchored to the inner shaft, except for the first, which is anchored to the housing

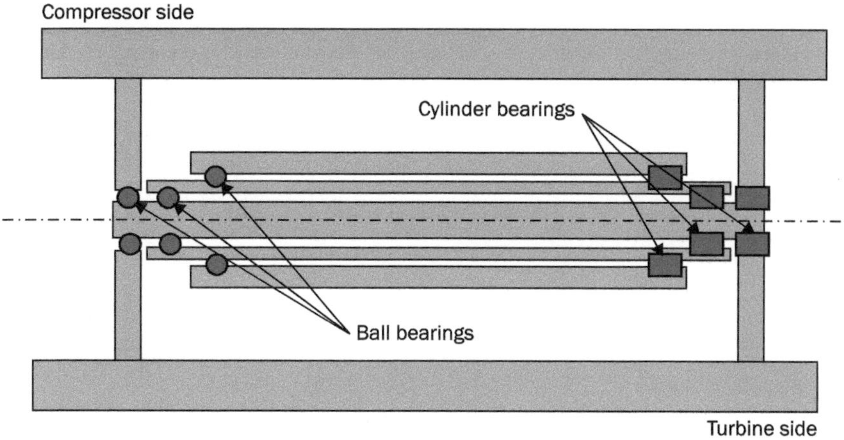

Table 3.4 Design matrix for the configuration shown in Figure 3.8

		Design parameters								
		Pair 1			Pair 2			Pair 3		
		Axial force	Radial force	Radial force	Axial force	Radial force	Radial force	Axial force	Radial force	Radial force
Functional requirements of shaft 1	Axial displacement	X								
	Radial displacement of point A		X							
	Radial displacement of point B			X						
Functional requirements of shaft 2	Axial displacement				X					
	Radial dicplacement of point A					X				
	Radial displacement of point B						X			
Functional requirements of shaft 3	Axial displacement							X		
	Radial displacement of point A								X	
	Radial displacement of point B									X

Figure 3.8 **Diagram showing a design solution in which each bearing is anchored to the case**

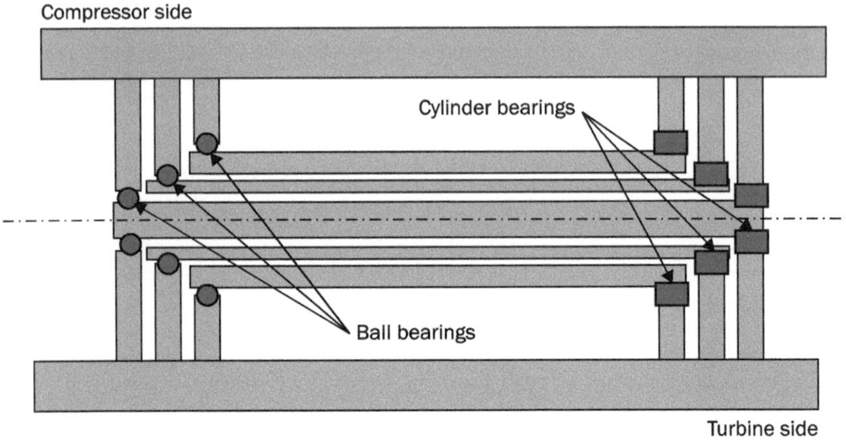

The physical configuration of this design matrix is the one in which each shaft is attached to the housing (and not to another shaft) by a different pair of bearings. A diagram of this solution is shown in Figure 3.8.

3.7.4 Application of the information axiom

There is still one ambiguity to be resolved: the ball bearings can be placed on the air intake side or the exhaust side. To resolve this, we will use the Information Axiom, which states that the probability of success must be maximized. New information must therefore be introduced in order to establish the probability of success, basically concerning reliability.

It is easy to argue that ball bearings are more stressed than cylinder bearings for two reasons: 1) they are under greater stress because they must support combined axial and radial loads, while cylinder bearings only support radial loads; and 2) they support a higher concentration of stresses because the geometry of a ball bearing generates point contacts on the rolling elements, while that of a cylinder bearing generates linear contacts. These are sufficient reasons to state that for a given lifetime and loads, the reliability of ball bearings will be lower.

On the other hand, because the mechanical properties of the materials (for example, hardness) decline with temperature, it is also easy to argue that the compressor side is a less aggressive environment than the turbine side. Consequently, to increase the probability of success, the elements under the most stress should be placed wherever the environment is less aggressive: the ball bearings must be placed on the compressor side. This is precisely the configuration used by Rolls-Royce for the high-pressure shaft on the turbofan shown in Figure 3.5.

3.7.5 Conclusions

Axiomatic Design is a design technique with a high added value because it reduces the number of possible initial designs to one while consuming almost no resources.[41] Thus, Axiomatic Design provides a conceptual design that satisfies the motivation. The configuration chosen is as close to the ideal design as the system permits.[42]

The selected configuration is the first one that must be studied in the detailed design. Consequently, if the detailed design shows that this solution is feasible, it saves all of the costs associated with the detailed design of the other configurations that would eventually be ruled out. If the detailed design finds deficiencies in the solution's compliance with the list of functional requirements and constraints for a shaft on a specific engine, the conceptual design must be modified slightly.[43]

3.8 Quantitative study of the design matrix

Because there can be no direct dependency between the functional requirements, it must be possible to vary one of them without changing the value of the others. This means that we can represent a set of r functional requirements as $y' \in \mathbb{R}^r$ using its coordinates in the canonical basis.[44] Let $\{y'_1, \ldots, y'_r\}$ be the canonical basis of $\mathbb{R}^r$. On the other hand, if we establish a solution characterized by a set of q design parameters that can be varied independently, the design parameters can be identified using the coordinates of a vector $x \in \mathbb{R}^q$ in the canonical basis $\{x'_1, \ldots, x'_q\}$. This solution results in an indirect dependency on the functional requirements. In an approach for small variations, this can be represented by design matrix A using the expression:

$$y' = Ax \tag{3.45}$$

The set $\{Ax'_1, \ldots, Ax'_q\}$ is a system generating the image of the linear map. In order to implement the design, it must be $q \geq r$.[45] Similarly, in order to implement the design, the rank of A must be r.[46] Because the rank of A is r, its row vectors $\{a_1^t, \ldots, a_r^t\}$ are linearly independent. In addition, because the rank of A is r, the vector set $\{Aa_1, \ldots, Aa_r\}$ is also linearly independent and is a basis of $\mathbb{R}^r$. Consequently, matrix AA^t, whose column vectors are the elements in set $\{Aa_1, \ldots, Aa_r\}$, also has rank r, and is therefore invertible. If the linear map is restricted to vectors $z = A^t e$ using $e \in \mathbb{R}^r$, which belong to the subspace engendered by the column vectors $\{a_1, \ldots, a_r\}$, we have $y' = Az = AA^t e$. Let the vectors $\{e_1, \ldots, e_r\}$ be the vectors that verify $y'_i = AA^t e_i$. Based on these, we can find the vectors $z_i = A^t e_i = A^t(AA^t)^{-1} y'_i$. The column vectors in the matrix $Z = A^t(AA^t)^{-1}$ are therefore the combination of design parameters that enable us to vary the functional requirements independently. We can check that $Z^t Z = (AA^t)^{-1}$ and that $AZ = I$, being I the identity matrix. The kernel of the linear map A is comprised of vectors $b \in \mathbb{R}^q$, which verify $Ab = 0$, i.e., are orthogonal to vector set $\{a_1, \ldots, a_r\}$. Let $\{b_{r+1}, \ldots, b_q\}$ be a basis of the kernel of the linear map, and B the matrix comprised of those column vectors. The design parameter vector could then be expressed as $x = Z\alpha + B\beta = A^t(AA^t)^{-1}\alpha + B\beta$, and the linear map as $y' = Ax = AZ\alpha + AB\beta = AA^t(AA^t)^{-1}\alpha = \alpha$, where $\alpha \in \mathbb{R}^r$ and $\beta \in \mathbb{R}^{q-r}$. Thus, we finally obtain $x = A^t(AA^t)^{-1}y' + B\beta$. Because $AB = 0$ is verified, the design parameter vector has the lowest norm when no vector from the kernel is added to it: $\|x\|^2 = x^t x = y'^t(AA^t)^{-1}y' + \beta^t B^t B\beta$. The result when $\beta = 0$ is a set of vectors x, which achieves the desired response y' by varying the greatest number of design parameters so that the norm for vector x is minimal. Based on vector $x_i = Zy'_i + B\beta_i$, the matrix $X = Z + B[\beta_1, \ldots, \beta_r]$ can be constructed. Its column vectors maintain the independence of the functional requirements for any value of vectors $\{\beta_1, \ldots, \beta_r\}$. A design parameter can be removed, for example x'_i, causing all column vectors in X to be orthogonal to vector x'_i. A maximum of $q - r$ design parameters can be eliminated; therefore, by choosing $q - r$ different vectors in the canonical basis $\{x'_1, \ldots, x'_q\}$, we can construct matrix X', whose column vectors are the vectors chosen for removal. The condition for eliminating the selected design parameters is $X'^t X = 0$, i.e., $X'^t Z + X'^t B[\beta_1, \ldots, \beta_r] = 0$. Because the ranks of X' and B are both $q - r$, the matrix $X'^t B$ can be invertible.[47] Therefore, $[\beta_1, \ldots, \beta_r] = -(X'^t B)^{-1} X'^t Z$. Finally, by replacing it in the expression of X, we obtain $X = Z - B(X'^t B)^{-1} X'^t Z = [I - B(X'^t B)^{-1} X'^t]Z$.

Definition: The *adjustment directions* are the column vectors in matrices[48]

$$Z = A^t(AA^t)^{-1} \tag{3.46}$$

$$X = Z + B[\beta_1, \ldots, \beta_r] \tag{3.47}$$

$$X = [I - B(X'^t B)^{-1} X'^t]Z \tag{3.48}$$

Matrix AA^t is diagonalizable because it is real and symmetrical, and can therefore be expressed as $AA^t = U \Lambda U^t$, where U is a matrix in which the columns are the eigenvectors of AA^t normalized to the unit, and Λ is a diagonal matrix where the elements of the diagonal are the eigenvalues placed in the same order as the eigenvectors. Because orthonormal eigenvectors can always be chosen for a symmetrical matrix, we have $U^tU = I$. Therefore, $(AA^t)^{-1} = U\Lambda^{-1}U^t$ is verified. We can therefore deduce that AA^t and $(AA^t)^{-1}$ have the same eigenvectors, while their eigenvalues are inverse to each other. On the other hand, all eigenvalues of AA^t are positive. Indeed, if $AA^tu_i = \lambda_iu_i$, then $0 \leq \|A^tu_i\|^2 = u_i{}^tAA^tu_i = \lambda_iu_i{}^tu_i = \lambda_i$. However, because the rank of the linear map is r, it must always be $A^tu_i \neq 0$. Consequently, it is $\lambda_i > 0$.

On the other hand, the maximum response is obtained when the norm of y' is maximum; in other words, when $\|y'\|^2 = y'^t\, y' = x^t\, A^t\, Ax = z^t\, A^tAz = e^t\, AA^t\, AA^te = e^t\, (AA^t)^2\, e$ is maximum. The directions that make the response extreme are obtained with the constraint $x^tx = 1$ or $e^te = 1$. The Lagrange multiplier method makes it possible to write the functions that must be extreme as $x^t\, A^tAx - \lambda_xx^tx$ and $e^t\, (AA^t)^2\, e - \lambda_ee^te$, resulting in the following eigenvalue problems: $A^tAx = \lambda_xx$ and $(AA^t)^2\, e = \lambda_ee$. From $AA^tu_i = \lambda_iu_i$, we can obtain $(AA^t)^2u_i = \lambda_iAA^tu_i = \lambda_i^2u_i$. AA^t and $(AA^t)^2$ therefore share the same eigenvectors. The eigenvalues in the second matrix are the squares of those in the first. From $AA^tu_i = \lambda_iu_i$, we obtain $A^tA(A^tu_i) = \lambda_iA^tu_i$. Therefore, the eigenvectors of A^tA that do not belong to the kernel of A are parallel to A^tu_i, and share the same eigenvalues as eigenvectors u_i in AA^t. The other eigenvalues of A^tA are null because their eigenvectors belong to the kernel of A. Note that if $u_i{}^tu_j = \delta_{ij}$, then $\|A^tu_i\|^2 = u_i{}^t\, AA^tu_i = \lambda_iu_i{}^tu_i = \lambda_i$ and the eigenvectors of A^tA that are not part of the kernel of A are $v_i = \lambda_i^{-1/2}\, A^tu_i$.

Because the chosen eigenvectors constitute an orthonormal basis, spectral decomposition of the identity $I = \sum_{i=1}^{r} u_iu_i{}^t$ can be used, alongwith $A = \sum_{i=1}^{r} u_iu_i{}^tA$, to find the different modes $A_i = u_iu_i{}^tA$ in the design matrix. This matrix can be written as $A = \sum_{i=1}^{r} A_i = \sum_{i=1}^{r} \lambda_i^{1/2} u_iv_i{}^t$. The modes in the matrix only respond to the corresponding eigenvector, i.e., $A_iv_j = u_iu_i{}^t\, AA^tu_j = \lambda_j\delta_{ij}u_i$. It is also easy to obtain the following relationships: $A_iA_j{}^t = \delta_{ij}\lambda_ju_iu_j{}^t$ and $AA^t = \sum_{i=1}^{r} \lambda_iu_iu_i{}^t$. When the eigenvalues are ordered from higher to lower, $\lambda_1 \geq \lambda_2 \geq \ldots \geq \lambda_r > 0$, the eigenvector $v_1 = \lambda_1^{-1/2}\, A^tu_1$ generates the response with maximum norm $y' = Av_1 = \lambda_1^{1/2}u_1$. The functional requirement with the greatest deviation therefore corresponds to that component of eigenvector u_1 which has the greatest absolute value. If there are n eigenvalues equal to the maximum eigenvalue, there will be n functional requirements with the maximum deviation.

Definition: The *dominant functional requirement* is the one that displays the greatest deviation. For every eigenvector in matrix AA^t associated with the maximum eigenvalue, the dominant functional requirement is the one whose order number matches the order number of the coordinate whose absolute value is highest in the eigenvector, i.e., if $|u_i|$ is the vector formed with the absolute values of the coordinates of eigenvector u_i, and $|u_i|_k$ is the kth coordinate of that vector, then y'_j is a dominant functional requirement if $\max_k\left(|u_i|_k\right)=|u_i|_j$ is verified.

3.8.1 Diagonalization

From this study, we can deduce that it is always possible to maintain the independence between requirements if we act on the design parameters by following the strategy of varying several at the same time as indicated by the column vectors in matrix Z. If we take the vectors of Z and B as a basis, the linear map takes the diagonal form $[I_{r\times r}|\ 0_{r\times(q-r)}]$. However, the need to act on the design parameters by following the directions $\{z_1,\ldots,z_r\}$ (or any variation caused by the addition of vectors from the kernel of A to expressions (3.46), (3.47) and (3.48)) means that on the next level of the design hierarchy, the functional requirements are vectors $\{x_1,\ldots,x_r\}$, and not the initial design parameters $\{x'_1,\ldots,x'_q\}$. This parameter change can also be written as a change in the linear map, so that the new linear map is $y' = AA^te$, where the design parameter set that makes the response independent is $\{e_1,\ldots,e_r\}$, whose components are the vectors in matrix $(AA^t)^{-1}$.[49] We can write that $e = (AA^t)^{-1}\ \alpha$, so that by taking vector α as new coordinates (new design parameters), the application becomes diagonal: $y' = I\alpha$.

Diagonalization Theorem: Any design matrix with maximum rank admits a set of design parameters that turn it into the identity matrix (square, diagonal and with non-null elements equal to one).

Proof: Simply replace the column vectors from (3.47) in (3.45).

Please note that this design parameter change must be implemented physically using some physical device without violating the constraints of the problem. When this solution is not feasible, independence can only be ensured by recreating the design matrix, for example, by seeking a new design point where the matrix is diagonal, or completely redoing the design from the beginning to find a solution completely different from the previous one.

Note that if the design matrix has been appropriately created, the absolute value of each component of the column vectors calculated according to expression (3.48)[50] must have a value less than or equal to 1.0, and the absolute value of at least one of the components of each vector must be equal to 1.0. In fact, when their value is over 1.0, the permitted margin of variation of the corresponding design parameter is exceeded, meaning that the response cannot be varied in the entire expected range.[51] Thus, the product either loses performance, or the tolerances used are too severe. If the absolute value of the design parameters in a design is less than 1.0, when the design parameters cover their entire permitted range, the response will exceed the acceptance limits (and violate the Information Axiom). For this reason, it is advisable for column vectors $X = [x_1, \ldots, x_r]$, obtained from (3.47) and (3.48), to have a maximum component with an absolute value close to 1.0. This means that a good indicator of the best combination of design parameters is

$$D = \sqrt{\sum_{j=1}^{r} \left(\max_i \left(\left| x_{ij} \right| \right) - 1 \right)^2} \qquad (3.49)$$

where $X = [x_1, \ldots, x_r] = \begin{pmatrix} x_{11} & \cdots & x_{1r} \\ \vdots & \ddots & \vdots \\ x_{q1} & \cdots & x_{qr} \end{pmatrix}$ is given by (3.48).

The chosen combination of parameters should have the value of D closest to zero.[52]

Property: An ideal design meets the condition $D = 0$.

Proof: If $D > 0$, we have one of the following two cases. Case 1: If $\max_i \left(\left| x_{ij} \right| \right) < 1$, then, if the parameters are varied throughout the permitted interval, the condition $\max_i \left(\left| x_{ij} \right| \right) = 1$ will be met. Consequently $|y'_j|$ will be greater than 1.0, exceeding the acceptance limit and failing to meet the Information Axiom. Case 2: If $\max_i \left(\left| x_{ij} \right| \right) > 1$, the necessary variation range for a design parameter exceeds the permitted range. Excessively strict tolerances are therefore being imposed. If the tolerances are too strict, Corollary 6 is not being met on the next level of the design hierarchy.

Property: All functional requirements of an ideal design are dominant.

Proof: The Information Theorem establishes that an ideal design has a square, diagonal design matrix. On the other hand, all elements on the diagonal have a value of 1.0 because, according to the previous property, $D = 0$ must be met. Furthermore, the Diagonalization Theorem states that this matrix exists. Consequently, all of its eigenvalues are 1 and its eigenvectors are the canonical basis of $\mathbb{R}^r$. All of the functional requirements are therefore dominant.

The Diagonalization Theorem states that the ideal design always exists, so it is the synthesis operator's job to find it. The bottleneck is usually the creative process. If the design team does not invent a solution with the ideal design matrix, the invention will not be the best.

3.8.2 Measuring the degree of independence

As we have seen, if the rank of the design matrix is maximum, the functional requirements can be kept independent by varying the design parameters in accordance with certain directions. However, in the presence of errors and noises, the dependency remains active. If several parameters must be varied at the same time, a small error may occur, slightly modifying the direction with respect to the adjustment direction that should be followed. This small deviation can act on another adjustment direction and couple the design. This effect is amplified if the adjustment directions are not orthogonal. In the previous section, it was proven that the scalar product of these vectors is $Z^t Z = (AA^t)^{-1}$. Thus, if the column vectors in Z are orthogonal to each other, the matrix AA^t must be diagonal. On the other hand, the vectors in Z use the greatest number of design parameters possible. This means that controlling a particular functional requirement may require all of the design parameters, and may therefore disturb the rest of the functional requirements with noises. This effect is measured by the modulus of the row vectors in matrix A, which are the square root of the elements on the diagonal in matrix AA^t. To decrease this effect, the greatest number of terms must be nulled in the design matrix. Nevertheless, it is necessary to ensure the presence of a dominant element in each row of A, which is achieved by normalizing the vectors using the component with the highest absolute value. This leads us to indicators I_1 and I_2 shown below. $I_1 \in [0,1]$ and $I_2 \in [0,1]$ are verified. The maximum independence is obtained when $I_1 = I_2 = 0$, and the maximum dependency when $I_1 = I_2 = 1$. The first indicator measures the importance of the terms outside of the main diagonal (orthogonality of the adjustment directions), and the second indicator measures the importance of the non-dominant terms (number of design parameters affecting a functional requirement).[53]

$$I_1 = \frac{2}{r(r-1)}\sum_{i=1}^{r-1}\sum_{j=i+1}^{r}\frac{\left|\sum_{k=1}^{q}A_{ik}A_{jk}\right|}{\sqrt{\sum_{k=1}^{q}A_{ik}^2}\sqrt{\sum_{k=1}^{q}A_{jk}^2}} \qquad (3.50)$$

$$I_2 = \frac{1}{r(q-1)}\sum_{i=1}^{r}\left(\frac{\sum_{k=1}^{q}\left|A_{ik}\right|}{\max_k\left(\left|A_{ik}\right|\right)}-1\right) \qquad (3.51)$$

Another way of measuring independence, introduced by Suh (2001: 155), is by using reangularity and semiangularity. Reangularity relates the angles between the axes of the design parameters, and semiangularity measures the magnitude of the diagonal elements. Reangularity is the absolute value of the product of the sines of all angles formed by taking the different design parameters two by two in the space of definition for the functional requirements (Suh, 1990: 116; Rinderle and Suh, 1982).

$$R = \prod_{i=1}^{n-1}\prod_{j=i+1}^{n}\sqrt{1-\frac{\left(\sum_{k=1}^{n}A_{ki}A_{kj}\right)^2}{\left(\sum_{k=1}^{n}A_{ki}^2\right)\left(\sum_{k=1}^{n}A_{kj}^2\right)}} \qquad (3.52)$$

$$S = \prod_{j=1}^{n}\left[1-\frac{\left|A_{jj}\right|}{\sqrt{\sum_{k=1}^{n}A_{kj}^2}}\right] \qquad (3.53)$$

Indicators R and S are defined so that when both are equal to 1.0, the design is uncoupled (diagonal matrix). When both are equal to each other, but not equal to one, the design approaches a quasi-coupled design (triangular matrix). In all other cases, the design is coupled, and both indicators are less than one. As Suh points out (2001: 156), R and S do not measure the alignment between the axes for the design parameters and functional requirements. Suh (1990: 139) and Su, Shie-Jie and Lin (2003) have studied other ways of measuring the degree of independence.

3.9 Example application: flow and temperature control[54]

The controlled supply of a hot liquid is a recurring problem in engineering. The problem is usually to find a regulating device capable of providing the appropriate temperature and flow. One possible solution to the problem is to mix two liquid flows at different temperatures: one cold and the other hot.

In an initial approach, the mass flow rate of a liquid through a certain section is reflected by the following expression: $\dot{m} = \sqrt{2\rho(P - P_0)}A$, where ρ is the density of the liquid (which we assume will change little with temperature), P is the supply pressure of the liquid, P_0 is the output pressure of the liquid, and A is the effective outlet area. In the solution studied, two flows are mixed to provide the desired mass flow rate, $\dot{m}$, at the desired temperature, T. Each flow has a different temperature, comes at a different pressure, and passes through a different area. If we identify the flows with subscripts 1 and 2, mass and energy conservation provides the output flow and temperature:

$$\dot{m} = \sqrt{2\rho(P_1 - P_0)}A_1 + \sqrt{2\rho(P_2 - P_0)}A_2 \tag{3.54}$$

$$T = \frac{\sqrt{P_1 - P_0}A_1T_1 + \sqrt{P_2 - P_0}A_2T_2}{\sqrt{P_1 - P_0}A_1 + \sqrt{P_2 - P_0}A_2} \tag{3.55}$$

On the other hand, a proper approach to the design problem requires specifying the variation ranges of the design parameters and the response. We will assume that the pressures cannot exceed a certain maximum pressure P_M, so that $P_1 \in [P_0, P_M]$ and $P_2 \in [P_0, P_M]$. Both areas can vary from 0 to a maximum value: $A_1 \in [0, A_M]$ and $A_2 \in [0, A_M]$. If we assume that T_1 is the cold temperature, the temperature of the other flow will be between the cold value and the maximum temperature permitted: $T_2 \in [T_1, T_M]$. The expected response must also be between a minimum value and a maximum value: $\dot{m} \in [0, \dot{m}_M]$ and $T \in [T_1, \chi T_M]$. These variation ranges enable us to establish the following encoded variables (see Eqs. (3.12) and (3.13)):

$$P_1 = \frac{P_M + P_0}{2} + x_1\frac{P_M - P_0}{2}, \qquad P_1 = \frac{P_M + P_0}{2} + x_2\frac{P_M - P_0}{2},$$

$$A_1 = \frac{A_M}{2} + x_3\frac{A_M}{2}, \quad A_2 = \frac{A_M}{2} + x_4\frac{A_M}{2}, \quad T_2 = \frac{T_M + T_1}{2} + x_5\frac{T_M - T_1}{2},$$

$$\dot{m} = \frac{\dot{m}_M}{2} + y_1 \frac{\dot{m}_M}{2} \quad \text{y} \quad T = \frac{\chi T_M + T_1}{2} + y_2 \frac{\chi T_M - T_1}{2} \tag{3.56}$$

Using these variables, equations (3.54) and (3.55) can be linearized around $x_i = 0$. The design equations in $x_i = 0$ provide the relationships for the center of the intervals:

$$\dot{m}_M = 2\sqrt{\rho(P_M - P_0)}A_M, \quad \chi T_M = \frac{T_M + T_1}{2} \tag{3.57}$$

These equations enable us to scale the response margins with the tolerances. As we can see, the output temperature will be roughly half of the maximum temperature. With the intervals centered, design equations (3.54) and (3.55) are transformed into:

$$\begin{Bmatrix} y_1 \\ y_2 \end{Bmatrix} = \begin{bmatrix} \frac{1}{4} & \frac{1}{4} & \frac{1}{2} & \frac{1}{2} & 0 \\ \frac{1}{4} & -\frac{1}{4} & \frac{1}{2} & -\frac{1}{2} & 1 \end{bmatrix} \begin{Bmatrix} x_1 \\ x_2 \\ x_3 \\ x_4 \\ x_5 \end{Bmatrix} \tag{3.58}$$

This design matrix ($I_1 = 0$, $I_2 = 0.438$) can be uncoupled using the following vectors (Eq. (3.46)):

$$Z^t = \begin{bmatrix} \frac{2}{5} & \frac{2}{5} & \frac{4}{5} & \frac{4}{5} & 0 \\ \frac{2}{13} & -\frac{2}{13} & \frac{4}{13} & -\frac{4}{13} & \frac{8}{13} \end{bmatrix} \tag{3.59}$$

The distance to the optimal point for these vectors is $D = 0.434$. This distance can be improved or worsened by taking the parameters two at a time (Eq. (3.48)):

1. $X''^t = \begin{bmatrix} 1 & 0 & 0 & 0 & 0 \\ 0 & 1 & 0 & 0 & 0 \\ 0 & 0 & 1 & 0 & 0 \end{bmatrix}, X^t = \begin{bmatrix} 0 & 0 & 0 & 2 & 1 \\ 0 & 0 & 0 & 0 & 1 \end{bmatrix}, D = 1$

2. $X''^t = \begin{bmatrix} 1 & 0 & 0 & 0 & 0 \\ 0 & 1 & 0 & 0 & 0 \\ 0 & 0 & 0 & 1 & 0 \end{bmatrix}, X^t = \begin{bmatrix} 0 & 0 & 2 & 0 & -1 \\ 0 & 0 & 0 & 0 & 1 \end{bmatrix}, D = 1$

3. $X''^t = \begin{bmatrix} 1 & 0 & 0 & 0 & 0 \\ 0 & 1 & 0 & 0 & 0 \\ 0 & 0 & 0 & 0 & 1 \end{bmatrix}, X^t = \begin{bmatrix} 0 & 0 & 1 & 1 & 0 \\ 0 & 0 & 1 & -1 & 0 \end{bmatrix}, D=0$

4. $X''^t = \begin{bmatrix} 1 & 0 & 0 & 0 & 0 \\ 0 & 0 & 1 & 0 & 0 \\ 0 & 0 & 0 & 1 & 0 \end{bmatrix}, X^t = \begin{bmatrix} 0 & 4 & 0 & 0 & 1 \\ 0 & 0 & 0 & 0 & 1 \end{bmatrix}, D=3$

5. $X''^t = \begin{bmatrix} 1 & 0 & 0 & 0 & 0 \\ 0 & 0 & 1 & 0 & 0 \\ 0 & 0 & 0 & 0 & 1 \end{bmatrix}$ This design is not valid because it produces a singular matrix.

6. $X''^t = \begin{bmatrix} 1 & 0 & 0 & 0 & 0 \\ 0 & 0 & 0 & 1 & 0 \\ 0 & 0 & 0 & 0 & 1 \end{bmatrix}, X^t = \begin{bmatrix} 0 & 2 & 1 & 0 & 0 \\ 0 & -2 & 1 & 0 & 0 \end{bmatrix}, D=1.414$

7. $X''^t = \begin{bmatrix} 0 & 1 & 0 & 0 & 0 \\ 0 & 0 & 1 & 0 & 0 \\ 0 & 0 & 0 & 1 & 0 \end{bmatrix}, X^t = \begin{bmatrix} 4 & 0 & 0 & 0 & -1 \\ 0 & 0 & 0 & 0 & 1 \end{bmatrix}, D=3$

8. $X''^t = \begin{bmatrix} 0 & 1 & 0 & 0 & 0 \\ 0 & 0 & 1 & 0 & 0 \\ 0 & 0 & 0 & 0 & 1 \end{bmatrix}, X^t = \begin{bmatrix} 2 & 0 & 0 & 1 & 0 \\ 2 & 0 & 0 & -1 & 0 \end{bmatrix}, D=1.414$

9. $X''^t = \begin{bmatrix} 0 & 1 & 0 & 0 & 0 \\ 0 & 0 & 0 & 1 & 0 \\ 0 & 0 & 0 & 0 & 1 \end{bmatrix}$ This design is not valid because it produces a singular matrix.

10. $X''^t = \begin{bmatrix} 0 & 0 & 1 & 0 & 0 \\ 0 & 0 & 0 & 1 & 0 \\ 0 & 0 & 0 & 0 & 1 \end{bmatrix}, X^t = \begin{bmatrix} 2 & 2 & 0 & 0 & 0 \\ 2 & -2 & 0 & 0 & 0 \end{bmatrix}, D=1.414$

Cases 5) and 9) result in a singular design matrix that makes it impossible to vary all of the functional requirements. Of the designs studied, the best is the third (D = 0), in which the input pressures and temperatures are kept fixed, and only the two areas are varied.

To accept the design, it is necessary to implement a physical system capable of creating the combination of areas according to matrix X in point 3). This matrix says that the area must be modified in such a way that if one area increases by one amount, the other must increase or decrease by the same amount. This can be implemented using a gate system in which the gate has two degrees of freedom, horizontal and

Figure 3.9 **Solution adopted to uncouple the liquid flow and temperature control problem**

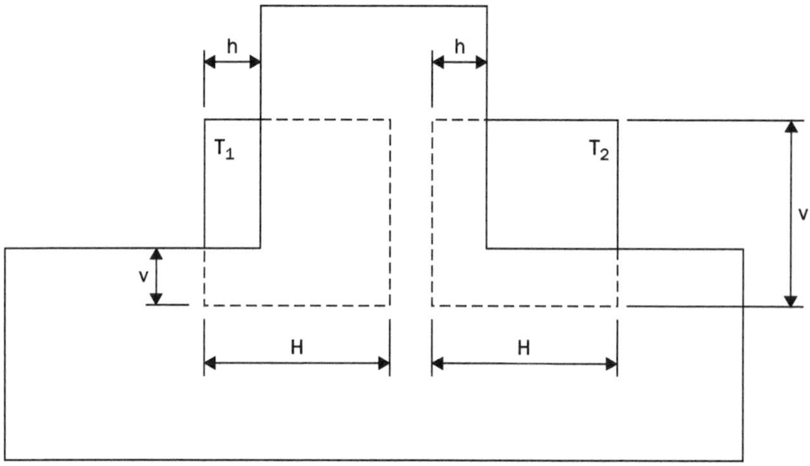

Note: The T-shape plate uncovers different portions of the rectangular fluid passage areas when it moves horizontally and vertically.

vertical. Figure 3.9 shows a diagram of this system, with a T-shaped plate and two rectangular input areas. The two areas can be expressed as follows:

$$A_1 = (V - v)h, \qquad A_2 = (V - v)(H - h) \tag{3.60}$$

The two new design parameters are the horizontal and vertical positions of the plate, which must verify: $h \in [0, H]$ and $v \in [0, V]$. On the other hand, the pressures and temperatures are no longer used as control variables, so their value is constant plus/minus a tolerance fixed by δ. Thus, the new design parameters are:

$$P_1 = (1 + x_1\delta)P_M, \qquad P_2 = (1 + x_2\delta)P_M,$$

$$v = \frac{V}{2} + x_3\frac{V}{2}, \qquad h = \frac{H}{2} + x_4\frac{H}{2} \quad \text{and} \quad T_2 = (1 + x_5\delta)T_M \tag{3.61}$$

The new equations for the center of the interval are:

$$\dot{m}_M = 2\sqrt{\rho(P_M - P_0)}VH, \qquad \chi = 1 \tag{3.62}$$

The maximum output temperature now matches the maximum temperature. Once the intervals are centered, the new design equations are:

$$\begin{Bmatrix} y_1 \\ y_2 \end{Bmatrix} = \begin{bmatrix} -\dfrac{\delta}{4}\dfrac{P_M}{P_M - P_0} & -\dfrac{\delta}{4}\dfrac{P_M}{P_M - P_0} & -1 & 0 & 0 \\ -\dfrac{\delta}{4}\dfrac{P_M}{P_M - P_0} & \dfrac{\delta}{4}\dfrac{P_M}{P_M - P_0} & 0 & 1 & \delta\dfrac{T_M}{T_M - T_1} \end{bmatrix} \begin{Bmatrix} x_1 \\ x_2 \\ x_3 \\ x_4 \\ x_5 \end{Bmatrix} \quad (3.63)$$

Due to the low value of the tolerances ($\delta \ll 1$), the resulting design matrix is practically diagonal ($I_1 = I_2 = 0$). The solution adopted is the 'single-lever faucet' concept. At this point, the final steps would be the detailed design and tolerance design for the two control parameters. To complete this example, it would be interesting for the reader to study those cases where the density changes with the temperature, and the temperature on the cold side is also variable.

3.10 Notes

1. In terms of the nomenclature introduced in Chapter 1, the object is the motivation and the list of needs that define it. As a minimum, this would be the information obtained in Step 1, as described in Section 1.9: Identifying the need.
2. Once this domain is reached, the customer's needs have been specified in terms of functional requirements and constraints. In order to move from the customer domain to the functional domain, at least Steps 2, 3 and 4 described in Section 1.9 must be performed. The result is a series of specifications. Suh (1990: 28–29) acknowledges that determining a good set of functional requirements based on fuzzily perceived and often poorly defined needs requires training, a thorough market study in many cases, and a large number of iterations. The use of the adjective 'functional' to describe this domain emphasizes that what is described in this domain is the response or expected output from the device (Suh, 1990: 38).
3. To satisfy the functional requirements, a series of possible preliminary solutions is created, which makes it possible to define the design parameters necessary to meet the functional requirements (system design). The adjective 'physical' reminds us that this domain includes everything expected for generating the desired output. The design must be optimized by choosing the design parameters appropriately during the preliminary and detailed design (parameter design) stages, and establishing their tolerances (tolerance design). To reach this domain, Steps 5, 6, 7 and 8 described in Section 1.9 must be performed. The transfer functions for the product are the work object.

4. Finally, to produce the product specified by the design parameters, we must develop a process characterized by the process variables (Steps 9 and 10 described in Section 1.9). The design process, in turn, includes all of the steps in a general design and, in particular, the system, parameter and tolerance design. The object is the transfer function for the process.
5. In addition to the spaces of definition proposed in Section 2.9 and the domains proposed by Axiomatic Design, other groupings and divisions can be made. For example, Huang (2002) proposes the design workspace and the review workspace.
6. In general, it is easier to reason using transfer functions that link the functional domain to the physical domain. All conclusions obtained will be equally valid for other transfer functions.
7. In Section 2.9, set *C* was divided into smaller subsets using tolerances. To avoid overloading the notation, the examples in this chapter will use set *C*, but are equally applicable to any element $(x_1)_{i1} \times \ldots \times (x_q)_{i_q}$ resulting from the partition of *C*.
8. This expression assumes that the requirements are statistically independent, and that they have a uniform probability distribution within the interval. As we will see, Axiomatic Design seeks independence among the requirements and achieves robust products. By definition, these are independent of the probability distribution adopted. For this reason, the above expression can be used as an initial approach. In the next chapter, Metric Design will correct the edge effects. Reliability-based design in Chapter 5 shows how to calculate the probability of success for a more general case.
9. Please note that there are many ways to adimensionalize a function. We can simply divide it by a characteristic value set by the transfer function itself, by physics, by an order of magnitude of interest to the problem, or by some other value with the same dimensions. For example, unlike the adimensionalization suggested here, Suh (1990: 119) proposes adimensionalizing the functional requirements with their desired value, rather than their range of variation. However, in the design process, the range of interest of a particular functional requirement is set by the customer: encoding (2.32) facilitates the assessment of the probability of success according to expression (3.1) because it explicitly includes the rejection and acceptance margins in the formulation of the design matrix. When this information is not available, we should use some other way of adimensionalizing the coordinated functions.
10. This normalization of the column vectors will be performed in the calculation of the reangularity and semiangularity.
11. As observed in (3.7), the design matrix depends on the acceptance limits in the functional requirements. This is essential because the proofs provided for some theorems (Sections 3.3 and 3.5) require the arbitrary variation of these limits. On the other hand, if a certain functional requirement is thought to be very important, its importance can be expressed by specifying the design range with no need to use a weighting factor (Suh, 2001: 44). In Chapter 4, cost will be included as a weighting factor.
12. Suh proposed a procedure for rearranging the rows and columns in the design matrix (Suh, 2001: 112; Suh, 1990: Appendix 10B.4). The procedure is as follows: 1) find a row with only one non-null element and reposition the rows

and columns so that the row and column with the non-null element appear first, 2) exclude the rows and columns that were already repositioned, and repeat Step 1 in the remaining submatrix until there are no more submatrices. This was the procedure used to find (3.10), starting with (3.9). Benavides and Garcia-Rodríguez (2011) present a modification of this algorithm. Lee (2006) uses graph theory to optimally reorganize the matrix. The Acclaro (2011) software by Axiomatic Design Solutions, Inc. makes it possible to rearrange the design matrix for square matrices on which the main diagonal is full.

13. Throughout this book, the same notation is used for intervals, vectors and real numbers. For example, y_1 can be the label associated with the first interval covered by variable y (notation used in Chapter 2), the first component of vector y, as shown in expression (3.11), or a vector where the values of the different components take certain values (Section 3.8).
14. Throughout this chapter, the dimensionless variables x_i and y_i are called design parameters and functional requirements, respectively. These variables are related to the design parameters m_i with dimensions and the functional requirements l_i, with dimensions as shown in (3.12) and (3.13).
15. Note that when the vector y_o is null, the intervals $f_i(\bigcup_j C_j)$ and $[\underline{l}_i, \overline{l}_i]$ are centered.
16. The idea that the acceptance interval can arbitrarily be made small might seem like a very restrictive hypothesis. However, this is not the case because, as we can see in the min shown in (3.20), its size is compared to the length of the response intervals. Thus, in relative terms, the acceptance interval must be very small compared to the response margin. Therefore, specifying that the acceptance intervals must be arbitrarily small is the same as specifying that the response margins must be arbitrarily large. In addition, the response margins can be made arbitrarily large if the tolerances or variation ranges of the design parameters are also made arbitrarily large. Large variations during the parameter design phase lead to robust products. For this reason, this theorem is of great interest.
17. An expression similar to (3.19) can be reached from design equation (3.11) if we accept that $y_i = \sum_{k=1}^{q} A_{ik} x_k$ are deviations from the functional requirements caused by the noises present in the design parameters. If all of the noises have a null mean and are statistically independent, we obtain $\sigma_{y_i} = \sqrt{E[y_i^2]} = \sqrt{\sum_{k=1}^{q} A_{ik}^2 \sigma_k^2}$, where σ_k is the standard deviation of design parameter x_k. The greater the number of design parameters, the greater the variation of the functional requirement. Note that $\sqrt{\sum_{k=1}^{q} A_{ik}^2} \leq \sum_{k=1}^{q} |A_{ik}|$ is always verified for any set of numbers, and that the maximum variation of a parameter always exceeds its standard deviation $\sigma_k < |x_k|$; therefore, (3.19) is a more severe restriction than would be obtained through statistical calculations. However, it is more useful because it is valid for both parameter and tolerance design. As we will see in Chapter 4, even if condition (3.24) is verified, the non-null terms outside the main diagonal can reduce product quality.

18. Theorem 8 of Axiomatic Design, which relates independence and tolerances (Suh, 1990: 122), states that: a design is uncoupled when the tolerance specified by the designer in a functional requirement is greater than the variation imposed by the design parameters that fall outside the main diagonal. For a coupled design, this condition is violated. Consequently, the probability of satisfying the specifications is lower (Suh, 1990: 168). This theorem is related to (3.24). However, as discussed in Section 3.8, the independence of the functional requirements (the uncoupling of the matrix) can be ensured by simply taking the values of the design parameters according to certain directions. What cannot be ensured when following these directions is that the design parameters will not exceed the permitted limits, will not cover the entire permitted limit, or will not be affected by noises. Axiom 2 and Theorem 8 ensure compliance with this last point. Thus, uncoupling does not require compliance with (3.24); it simply requires that the design matrix have the maximum rank.
19. Knowing the dominant elements makes it possible to turn the quantitative design matrices (3.7) and (3.8) into qualitative matrices (3.9) and (3.10).
20. As we saw in Chapter 2, the design process includes more aspects than the transfer function itself. It also involves generating the spaces of definition, and reducing the entropy of the probability distributions associated with them.
21. By definition, a set of functional requirements cannot contain functions that depend on each other. Suh (1990: 38) describes a set of functional requirements that depend on each other as redundant. In this book, in accordance with the nomenclature introduced in Chapter 1, it will generally be referred to as a list of needs.
22. Chapter 2 showed how the space of definition for the needs is related to the space of definition for the solution through the spaces of definition for the response and satisfaction.
23. This type of dependency originates from the element definitions or the laws of physics. For example, power, torque and angular speed are related by their definition regardless of the physical system providing the power, torque or rotation speed. If the torque varies, the angular speed or power will inevitably have to change.
24. This type of dependency originates from the configuration and arrangement of the different elements comprising the solution. If the solution adopted for delivering hot water is to mix a cold water flow with a variable flow of hot water at a fixed temperature, the temperature will also vary when the flow is modified by closing the valve.
25. This definition only requires checking that there are no direct dependencies. The Independence Axiom will also seek that there be no indirect dependencies. Finally, the Information Axiom will limit the number of parameters that affect the functional requirements. Axiomatic Design therefore postulates that a situation is better the smaller the number of dependencies affecting the requirements that define it.
26. It is usually better to use cost as a constraint rather than a functional requirement because cost is normally affected by changes in the other functional requirements. On the other hand, changes in the design parameters also tend to affect the cost, so that cost cannot be independent of the other

functional requirements in an uncoupled design. With cost as a constraint, the design is acceptable as long as it does not exceed the cost limit imposed. If cost were required as a functional requirement, the design would be coupled (Suh, 2001: 21).

27. As discussed in Section 3.8, the design matrix of an ideal design is the identity matrix. The nomenclature employed by Suh has been used here instead of the terms presented in Chapter 2 because they are equivalent. The system range is the interval generated by the response margins, and the design range is the interval generated by the acceptance limits. An equivalent statement could therefore be: 'This ensures that the response margins fall within the acceptance limits.' This is precisely the information reflected by the probability of an interval as defined in Section 2.10, Equation (2.36).
28. This theorem is a necessary, but not sufficient, condition for finding the best design. The definition of the best design must also include the Information Axiom.
29. If the design matrix cannot be converted to a diagonal because the designer cannot find a physical realization that implements the new functional requirements and design parameters, Suh (1990: 56) relaxes the criterion to accept that a triangular matrix (for example, a lower triangular matrix) satisfies Axiom 1. These solutions are described as decoupled or quasi-coupled. The argument is that a particular execution order of the design parameters makes it possible to modify the different functional requirements independently. However, strictly speaking, a quasi-coupled design violates Axiom 1 because it establishes an order in the functional requirements: this relationship of order is a dependency of each functional requirement on the requirements preceding them. To break this relationship of order, it is necessary to determine the directions by which the independence is maintained. For example, the lower triangular matrix $A=\begin{bmatrix} 1 & 0 & 0 \\ 1 & 1 & 0 \\ 1 & 1 & 1 \end{bmatrix}$ is uncoupled according to the adjustment directions given by the column vectors in $X=\begin{bmatrix} 1 & 0 & 0 \\ -1 & 1 & 0 \\ 0 & -1 & 1 \end{bmatrix}$ (see section 3.8).

    Nonetheless, if the design matrix is triangular, its rank is always maximum. Therefore, if the design matrix is only qualitatively known according to (3.9), it is useful to try to convert it to a triangular matrix because this will identify which elements could reduce the rank.
30. Volume V in expression (3.28) is obtained by finding the Cartesian product of the acceptance intervals (see Section 2.9) for each of the functional requirements.
31. If expression (3.1) is not valid for a given design, this is because the design was not good, and should have been discarded. This shows that it is wise to give Axiom 1 priority over Axiom 2 because otherwise the information content calculation would be extraordinarily complex. Nonetheless, Chapter 5 shows how to perform an approximate calculation using expression (3.28) for a general case.
32. The more design parameters affect a functional requirement, the more the requirement will vary due to variations in the design parameters. For this

reason, Suh (1990: 48–49) maintains that a quasi-coupled design (triangular matrix) can be worse than an uncoupled design (diagonal matrix); the former may present a higher information content. This ambiguity, if present, is resolved by Axiom 2: There are more occasions when a diagonal matrix can have a greater probability of success than a triangular one (see Equations (3.30) and (3.31)). Moreover, strictly speaking, a quasi-coupled design violates Axiom 1. (see note 29).

33. It is an interesting exercise to prove that if the mean and standard deviation of a distribution are fixed, the normal distribution is the continuous distribution defined all along the real line with maximum differential entropy.
34. Please note that in this proof no constraints were imposed on the *g* functions or the design parameters. In a real design situation, constraints could prevent the information content from being reduced as desired. In addition, the minimum information content is zero for both coupled and uncoupled designs. In this situation, without the Independence Axiom, it would be impossible to choose between the two.
35. The original numbering used by N. P. Suh has been respected to avoid confusion. However, some statements have been slightly modified. The initial formulations and original proofs can be consulted in the references. Corollary 2(*), which was not in the references, was also added to facilitate some proofs.
36. Functional coupling must not be confused with physical coupling, which might be beneficial for meeting Axiom 2 (Suh, 1990: 50). Two separate functions can be satisfied by a single part without functional coupling (Suh, 1990: 51). This might make it impossible to perform some functions simultaneously.
37. A design that is uncoupled in a particular operating range could be coupled in another working regime if the design is not linear. Chapter 6 will prove that linear designs are better than non-linear ones.
38. Suh (2001: 68) specifies that deviations should be eliminated, variations should be reduced, and large variations in the design parameters and process variables should be tolerated.
39. Santos et al. (2009) present a similar application, where Axiom 1 yields to a minimum number of external constraints at each one of the system's components.
40. This list of functional requirements was obtained for an infinitely rigid shaft. A real shaft has a particular bending rigidity that could require more than one position control point over the shaft. The introduction of this third point (point C) would break the independence of the functional requirements. Axiomatic Design would therefore recommend increasing the rigidity of the shaft until the third control point can be removed from the list of needs. This way, 'sufficient bending rigidity of the shaft' would appear as an additional constraint. For very long shafts, this constraint could enter into conflict with others. For example, for an aeronautical application, the minimum mass constraint could require the inclusion of a third support point on the shaft in order to increase the guiding accuracy of the shaft.
41. The number of initial configurations grows with the system size following an exponential growth. With N shafts and n bearings per shaft, we have the

following constraints: 1) at least two bearings in the configuration must be attached to the housing, and 2) at least one bearing on each shaft must be a ball bearing. There are N ball bearings that can be located near the compressor or turbine, $N(n-1)$ that can be ball or cylinder bearings, and $(N-1)n$ bearings that can be attached to the housing or inner shaft. This makes a total of $2^{N(2n-1)-n+1}$ initial configurations. With three shafts and two bearings per shaft, there are 256 initial configurations!

42. Other solutions could be found by applying other procedures, but the axioms warn that they would be worse. For example, physics would recommend mounting more than one bearing per shaft, as far apart as possible, in order to absorb the torques caused by the motion of the vehicle, and for at least one of the two bearings on each shaft to be a ball bearing in order to absorb axial forces. However, physics does not define a single possible design solution. The excess degrees of freedom are due to the following: 1) Physics defines the minimum number of bearings that must be used, but not the maximum. 2) Physics says nothing about the type of bearing, except that at least one on each shaft must be a ball bearing. 3) Physics says nothing about whether the bearing that holds one end of a shaft should be connected to another shaft or to the engine housing. A new criterion based on the cost of the solution would state that the number of elements must be minimal and that the largest number of identical parts possible should be used in order to reduce the number of different parts in stock, or to make replacement parts interchangeable. If we follow this line of reasoning, the solution would be two identical bearings per shaft (both ball bearings). This criterion would lead to the same design as Corollaries 3 and 5 used in isolation, which require reducing the number of parts to a minimum and using the largest number of symmetries possible.
43. Obviously, the fact that Axiomatic Design marks a solution as optimum does not prevent the industry from using others. If the list of constraints for a shaft on an engine were different from the one proposed, other solutions different from the ones obtained in this section could be better. Therefore, comparing solutions in order to catalog them as better or worse only makes sense when both solutions share the same set of functional requirements and constraints. For example, the Rolls-Royce Trent 500 (Airbus A340–500 and –600) uses this bearing configuration for the high-pressure shaft, but includes a third support point, a cylinder bearing, on the front of both the intermediate shaft and the low-pressure shaft. Another peculiarity is that the ball bearing on the low-pressure shaft is not supported by the housing, but by the intermediate shaft. The CFM56-5C engine (Airbus 340–200 and –300) by CFM International has only two shafts with a cylinder bearing between the shafts. Although these engines do not use exactly the same configuration selected by Axiomatic Design, they come very close to the ideal design from an Axiomatic Design perspective. For example, as we can see in Table 3, the elimination of one shaft reduces the number of crosses outside of the diagonal from 6 to 3. The placement of only the cylinder bearing between the shafts reduces the number of conflicting crosses from 3 to 1.
44. Vector y' is used instead of $y - y_o$ in (3.11). The apostrophe indicates that the functional requirements vector used in this section is a deviation with respect

to the midpoint of the acceptance interval. If the acceptance and response intervals were centered, y_o would be null.

45. If it were $q < r$, there would be an element of the canonical basis of $\mathbb{R}^r$, for example y'_r, that would not belong to the subspace engendered by $\{Ax'_1, \ldots, Ax'_q\}$. Under these conditions, the design parameter set either would not allow that functional requirement to vary, or it would not vary independently of the other functional requirements.
46. If the rank of the matrix were less than r, the generating system $\{Ax'_1, \ldots, Ax'_q\}$ would not have r linearly independent elements. Consequently, there would be an element in the canonical basis of $\mathbb{R}^r$, for example y'_r, that would not belong to that subspace.
47. The fact that the ranks for both matrices are maximum does not guarantee that the rank for the product will be. See case $\begin{bmatrix} 1 & 0 & 0 \\ 0 & 0 & 1 \end{bmatrix} \begin{bmatrix} 1 & 1 \\ 1 & 2 \\ 1 & 1 \end{bmatrix} = \begin{bmatrix} 1 & 1 \\ 1 & 1 \end{bmatrix}$.

    If the resulting matrix $X'^t B$ is not invertible, removing the design parameters chosen in matrix X' is not possible because it leaves some functional requirement unsatisfied (the matrix obtained using zeros to replace the columns in A associated with the eliminated design parameters has a rank lower than r).
48. The adjustment directions establish the proportion by which the design parameters must be varied so that the functional requirements vary independently. The adjustment directions in the form (3.46), (3.47) and (3.48) are the main result of this section and have been defined by the author in order to give a procedure for converting a coupled design matrix (not necessarily square) into an uncoupled square design matrix. The Diagonalization Theorem in section 3.8.1 enssures its existence. An illustrative example is given in section 3.9.
49. These vectors could be expressed as a linear combination of eigenvectors $\{u_1, \ldots, u_r\}$, but the direction marked by an eigenvector does not generally maintain the independence of the functional requirements.
50. When $X'^t B$ is not invertible, we can always use expression (3.47) with any value of the vectors $\{\beta_1, \ldots, \beta_r\}$.
51. In some situations, we can compensate for this by modifying another design parameter, but this would take us farther away from the adjustment direction.
52. The condition $D = 0$ is advisable during the parameter design phase. During the tolerance design phase, care must be taken with condition (3.23).
53. Note the similarity between the second indicator and expression (3.23). On the other hand, the first indicator is equal to $1 - \pi_o$, where π_o is the orthogonality index defined by Suh (1990: 139).
54. This example was used by N. P. Suh (2001: 119, 'Hot and cold water faucet') to illustrate Axiomatic Design procedures in a qualitative form. In this Section, a quantitative approach has been taken. The resulting design, which has one T-shaped moving plate (Figure 3.9), is slightly different from the ones described in the reference.

4

Metric design

Abstract: Metric Design studies the influence of noise and cost on the quality of the product designed. It establishes that the best product is the one with the lowest cost and the most predictable performance. This clashes with the ideas of some designers, who believe that their job is to design by optimizing the product's features, appearance, reliability, etc., and that the production engineer's task is to design a process capable of manufacturing that product (Taguchi, Elsayed and Hsiang 1989). In fact, both designs are intimately related (concurrent engineering), which is why it is not only important to determine the optimum value of the product's parameters, but also its tolerances. Obviously, not all conceptual designs will lead to the best product. This chapter examines some of the implications of Metric Design for the design process.

Key words: robustness, tolerance, noise, cost.

4.1 Introduction to metric design

Axiomatic Design, described in the previous chapter, views what is right in engineering as that which verifies the axioms to the greatest extent: excess dependencies and an excess information content lead to bad designs. However, two effects that are not directly included in the axioms are closely related to the satisfaction of the needs: cost and variability. In Axiomatic Design, cost is usually introduced as a constraint, while variability is only considered for its effect on information content.[1] If the information content is null, there is no way to distinguish between a product whose response is in the center of the acceptance interval and one that is right on one of the limits. In order to incorporate this difference, some of the proofs in the previous chapter imposed an arbitrary variation of the acceptance limits. The underlying idea is that not all points within the acceptance interval produce the same degree of satisfaction.[2]

The approach for correcting those differences described in this chapter is Metric Design, which regards a product's 'quality loss' as a measure of how 'bad' it is. Metric Design is based on the following main concepts:

- Quality loss implies an economic loss. It can therefore be defined in economic terms.[3]
- Variability constitutes a quality loss.

4.2 Quality loss

The economic losses imposed on society must be measured by the actual gap between the response of the product sold and its specifications. The acceptance limits of these functional requirements determine whether or not a product should be rejected.[4] The economic losses imposed on society cannot be evaluated solely by the level of rejections (defective parts) because such rejections should not reach the consumer. When defective products are not placed on the market, the consumer is not directly affected, except by the higher cost of the product. Rejections are not a matter of quality, but of cost (Taguchi, Elsayed and Hsiang, 1989). According to Taguchi, Elsayed and Hsiang (1989), quality loss is the economic loss borne by society from the time a product is placed on the market.

Definition: The *quality loss* is the economic loss borne by the entire society due to a production appearing on the market.[5]

This definition assumes that: 1) The entire society is sensitive to the quality of a product. 2) The measure of a product's quality is the economic impact that it produces on society, so that a greater impact indicates lower quality. 3) Our activities, in economic terms, always produce an economic loss for society. This definition is important for the following reasons:

1. Because it is defined as a loss, it has a lower bound. This makes it possible to establish extremes.

2. Defining it in economic terms gives it universal features. The units of measure for quality loss are monetary units, such as € or $, which can be understood and compared in any sphere of society: design firms, boards of directors, users, etc.
3. The economic impact measured affects the entire society, so that the definition is valid regardless of the particular agents involved, i.e. regardless of the specific company, customer, product or situation being analyzed.[6]
4. It means acknowledging that our actions, when observed globally, are inevitably imperfect, and inevitably involve an opportunity cost.

As mentioned in the introduction, variability constitutes a quality loss. In other words, a product whose response is near the extremes of the acceptance interval displays a greater quality loss than one whose response is in the center of the acceptance interval.[7] There are several arguments supporting the accuracy of this statement. One states that if the customer did not consider those products whose response is far from the center of the acceptance interval to be worse, he would not have centered the interval on that point. Another argument suggests that the probability that a product, subject to noise and inaccuracies, will produce a response outside the acceptance interval is greater if that product's response was initially near an extreme of the interval. A final argument explains that if the customer specified an acceptance interval, this is because a product with a response outside the acceptance interval generates dissatisfaction to the point that it must be replaced or disposed of, with the resulting economic loss. Metric Design therefore accepts the following postulate:

Metric Design Postulate: For a given production, the lowest quality loss is obtained with the product whose response coincides with the centers of the acceptance intervals.[8]

The design philosophy based on minimizing quality loss is Metric Design. Because the product obtained when designing with this methodology is the one displaying the lowest quality loss, it can be argued that the product obtained has the highest quality.[9]

4.2.1 Design for maximum quality

To determine the quality loss, we need to know whether or not a product is satisfactory for the customer. Customer satisfaction is measured by compliance with functional requirements, which are therefore the true indicators of product quality. The customer observes variations in performance over time and from one product to another with respect to the nominal values of the functional requirements. However, the customer buys hoping that the product will perform in line with the nominal values of the functional requirements throughout its lifetime. Variations in performance are perceived by the customer as a quality loss. Variation is therefore a phenomenon that must be taken into account while designing. When designing for maximum quality, minimum variation is always a need to consider.[10] In the long term, this design philosophy will have an effect on the company's profits. Above all, it will achieve a minimum quality loss, which means that the economic loss perceived by society will also be minimum. This way, the company will remain a market leader due to the quality of its product.[11] The objective of Metric Design is to optimize the final product, choosing the best architecture and the best nominal values and tolerances in order to achieve maximum quality (Ross, 1988; Taguchi, Elsayed and Hsiang, 1989).

Reducing the quality loss requires reducing cost and variability. Basically, the total cost associated with a product must be broken down as follows:

- Raw materials, energy, work and structural loads associated with proper production.
- Analogous resources wasted on useless production.
- Harmful side effects caused during the manufacture, use and life of the product.

On the other hand, the different factors affecting variation are:

- Controllable factors or parameters: These are directly controllable by the designer to minimize variation. They are divided into:
  - Control factors: The designer can choose their value freely without appreciably affecting manufacturing costs.
  - Tolerance factors: These are parameters that do affect manufacturing costs. When a tolerance is reduced, the cost increases, particularly if process capacity is poor.
- Noise factors: These cannot be controlled by the designer, and their value can change from unit to unit, from one environment to another, etc. Some of these are shown in Table 4.1.

Table 4.1 **Sources of noise**

NOISE	EXTERNAL	Environmental causes	Temperature, pressure, humidity, dust, electrical power, etc.
		Product use	Workload, frequency of use, intensity of use, human error during use, etc.
	INTERNAL	Performance degradation	Wear and tear, loss of strength in springs, etc.
		Production imperfections	Wear and tear, looseness, unstable processes, variations in raw materials, variations in machinery, lack of control, etc.

4.2.1.1 Stages of design

Reducing cost and variability requires an adequate design for both the product and the process.

- *Product design*: This is the most effective activity for reducing the effects of all types of noise. If the product is made insensitive to environmental conditions, operator changes, etc., the operating cost will be low. If it is also made insensitive to changes in raw materials and machinery, etc., the manufacturing cost will be low, and it will be possible to design less complex or less controlled processes.
- *Process design*: This activity reduces the process cost and the variations from one unit to another. The better the product design, the more these will be reduced. As we have seen, environmental factors, deterioration, and imperfections in the manufacturing process are possible sources of variation in the product characteristics with respect to their nominal value. The product design engineer deals with the first two sources. However, if the product design is robust enough, this will facilitate the work of the production engineers because the product characteristics will be highly insensitive to variations in production.

According to Taguchi, Elsayed and Hsiang (1989), both product design and process design must be divided into three phases.

- *System design*: During this phase, the architectures and technologies that could achieve the desired function are examined. The most appropriate one is selected in order to obtain a basic, functional prototype. During this stage, the experience and creativity of the designer or design team play an important role. Knowledge of the customer's needs and the manufacturing environment of the chosen technology is essential. It is during this phase that innovation takes place.

- *Parameter design*: The best values for the control parameters are determined, which are those values that result in the minimum quality loss. The value of the response must be as close to the nominal value as possible. Large variations must be set in advance for the noise factors, and low-performance materials and components must be chosen to minimize cost. If the response falls within the specifications, a product and/or process has been achieved at the minimum price, and tolerancing is not necessary. However, if the quality loss for the customer must be reduced, we must proceed to tolerance design.
- *Tolerance design*: The tolerances around the nominal setting for the control factors identified and studied during the parameter design phase are determined. Very strict tolerances increase the production cost, and broad tolerances excessively increase variations in performance; therefore, both extremes increase the quality loss. For this reason, we must use well-founded reasoning to choose tolerances.

4.2.1.2 Procedures for reducing variability

The procedures available to companies for reducing variability are:

- *Design quality (off-line quality)*. This takes place during the design process for the product and process. This phase achieves middle-term and long-term improvements in quality. It originated in Japan, and was implemented in the U.S. and Europe starting in 1980. It is at the heart of Total Quality Management. The tool used is DOE (Design of Experiments).[12]
- *Day-to-day quality (on-line quality)*. This takes place during the real-time production process. This phase achieves short-term improvements in quality. The tools used are SPC (Statistical Process Control) and SQC (Statistical Quality Control).

With adequate product and process design, day-to-day control will be minimal, but necessary for combating differences from one unit to another and detecting problems in the process. The main actions for ensuring on-line quality are:

1. **Preventive control:** The objective is to detect any deviations from nominal values that would result in rejections before they occur, in order to prevent and correct them. SPC (Statistical Process Control) is used for this purpose. It detects machine, operator and raw material

failures, wear and tear, etc. This represents an additional cost that must be compensated by a reduction in the quality loss for the customer. It controls day-to-day operations.

2. **Adaptive control:** Forward error information feedback is used to try to adapt the rest of the production line to the error in order to avoid rejection. This does not represent a major additional cost, and does not increase the quality loss for the customer.
3. **100 per cent final control:** This prevents defective units from reaching the customer. It is essential when the processes are not capable, as rejections will inevitably occur. The method used is SQC (Statistical Quality Control). This is based on inspection (defective products must be detected). This is the least economical method, as it keeps the response within the specifications, but increases the costs (inspection costs plus the cost of producing and/or repairing defective units).

The relationship between quality, variation, and design and production processes is shown in Table 4.2. Table 4.3 shows the relationship between quality, time, cost and the activity employed to maximize quality.

Table 4.2 **Reducing variation by the different activities in the configuration of a product**

TIME	AREAS	STAGES	REDUCTION IN THE VARIATION		
			ENVIRONMENTAL	DEGRADATION	UNIT TO UNIT
MIDDLE TERM, OFF-LINE	PRODUCT DESIGN	System design Parameter design Tolerance design	YES YES YES/NO*	YES YES YES	YES YES YES
	PROCESS DESIGN	System design Parameter design Tolerance design	NO NO NO	NO NO NO	YES YES YES
DAY TO DAY, ON-LINE	PRODUCTION	Preventive control Adaptive control 100% final control	NO NO NO	NO NO NO	YES YES YES

*YES/NO means that it is possible, but not advisable.
Source: Adapted from Taguchi, Elsayed and Hsiang (1989).

Table 4.3 Relationship between quality and cost for the different variation control activities

TIME	ACTIVITY	TIMING	COST	QUALITY
Past	SQC	After the process	HIGH	LOW
Present	SPC	During the process	MEDIUM	MEDIUM
Future	DOE	Before the process	LOW	HIGH

A need therefore arises to be able to measure the quality loss in different situations in order to distinguish the optimum design solution from all the others. This will be done, in a general manner, in the following section.

4.2.2 Quality loss function

The Metric Design approach postulates that the use of a product by a customer produces a quality loss that depends on that product's particular usage characteristics.

Definition: For an N-sized production, the quality loss is the sum of the quality loss for each product (specimen) comprising the production:

$$L_N = \sum_{j=1}^{N} L(y_j) \tag{4.1}$$

$$L{:}\mathbb{R}^r \to \mathbb{R},\ y \mapsto L(y) \tag{4.2}$$

The argument in the quality loss function (4.2) is the functional requirement vector whose components are the dimensionless variables defined in (2.32) and (3.13).[13] Subscript j in (4.1) indicates that the functional requirements vector is particularized for product number j, i.e. is the concrete response of specimen number j. This definition imposes: 1) that the quality loss is additive, and 2) that the function that enables us to calculate it is identical for each product.

For example, imagine that you want to shoot a jet of water at an object located at the same height as the nozzle, and nothing except that object should get wet. Thus, the water falling before position $\underline{l}$ or after position $\overline{l}$ produces insatisfaction. The functional requirement that describes the need is the distance (adimensionalized using $(\overline{l} - \underline{l})/2$) from a particle of water to the center of the object (whose position is $l_o = (\bar{l} + \underline{l})/2$). If the jet of water is shot at velocity V and inclination α, the argument that would appear in (4.2) is $y = 2(V^2 \sin(2\alpha)/g - l_o)/(\bar{l} - \underline{l})$, where g is the acceleration of gravity. When the experiment is repeated several times, the exact position of the nozzle with respect to the object is not known, so that variable l_o changes with each assembly. Similarly, the exit velocity and angle change with each assembly. These variables also change for each fluid particle, depending on the exit position in the nozzle mouth. Thus, the argument that would appear in (4.1) is $y_j = 2(V_j^2 \sin(2\alpha_j)/g - (l_o)_j)/(\bar{l} - \underline{l})$, where subscript j indicates a hose outlet region in a particular assembly. In this problem, specimen j is the water shot by each area of the outlet nozzle in each assembly, and the production is all water shot in all of the assemblies performed. The above theoretical model does not consider errors and noise due to interferences, leaks and splashes. To take into account all sources of variation, a series of experiments must be performed in which the distance is measured from the different volumes of water to a reference position. If l is that distance, then $y = 2(l - l_o)/(\bar{l} - \underline{l})$ is the quality indicator in (4.2) and $y_j = 2(l_j - l_o)/(\bar{l} - \underline{l})$ is the quality indicator for volume j in (4.1). In general, the specific choice of specimens within the production is an important step that depends on the design problem, and must be performed carefully.

Theorem: The quality loss function for a set of functional requirements (that can be developed in a Taylor series up to second-order terms) is:[14]

$$L(y) = C_o + (My)^t My \qquad (4.3)$$

$$M = \begin{pmatrix} \sqrt{C_1} & & 0 \\ & \ddots & \\ 0 & & \sqrt{C_r} \end{pmatrix} \qquad (4.4)$$

where C_o is the cost associated with purchasing a product, and M is a diagonal matrix whose elements are the square root of cost C_i, which assumes that the functional requirement associated with component i of vector y does not fall within the acceptance interval.[15]

Proof: The Metric Design postulate establishes $L(y) \geq L(0)$ for any vector $y \neq 0$ calculated with the dimensionless variables defined in (3.13).[16] Therefore, this must be $\left.\dfrac{\partial L}{\partial y_i}\right|_{y=0} = 0 \quad i = 1,\ldots,r$, and matrix $B = \dfrac{1}{2}\left[\left.\dfrac{\partial^2 L}{\partial y_i \partial y_j}\right|_{y=0}\right]$ composed of the second derivatives of the quality loss function cannot have any negative eigenvalues. On the other hand, the eigenvectors of B indicate the directions where the quality loss becomes extreme under the condition $y^t y = 1$. If matrix B were not diagonal, it would have at least two different eigenvalues.[17] In the direction of the eigenvector associated with the lowest eigenvalue, the quality loss would be lower than in the other direction. This privileged direction would cause the functional requirements not to be independent. Because functional requirements are independent by definition, matrix B must be diagonal. Finally, if any of the functional requirements reaches the value ± 1; in other words, if the functional requirement vector takes the form $y^t = (0,\ldots,0,\pm 1,0,\ldots,0)$, then one of the functional requirements has reached an extreme of the acceptance interval. Consequently, the corresponding element on the diagonal in matrix B must be the cost associated with such non-compliance. To finish this proof, we simply take $M^2 = B$, which always exists because all eigenvalues of B are non-negative.

As we can see, what the quality loss function does is weight each functional requirement with the square root of the extra cost produced by not satisfying that functional requirement. As mentioned in Chapter 3, Metric Design uses the cost structure to weight the relative importance of each functional requirement. On the other hand, expression (4.3) states that the quality loss function is, in an initial approach, a positive definite quadratic form (if no functional requirement is completely negligible, it must be $C_i > 0$ for every i). From a mathematical perspective, this is a function that provides the distance to the optimum product. Because this measure establishes that the points near the extremes of the intervals are less desirable than the inner points, this metric takes into account the edge effects.

$$L_N = \sum_{j=1}^{N} L(y_j) = \sum_{j=1}^{N} \left[C_o + (My_j)^t My_j \right] = NC_o + \sum_{j=1}^{N} y_j{}^t M^2 y_j \tag{4.5}$$

The variation of the response can be written as the variation around the average response:

$$\begin{aligned} L_N &= NC_o + \sum_{j=1}^{N} y_j{}^t M^2 y_j \\ &= NC_o + N\left(\frac{1}{N}\sum_{j=1}^{N} y_j\right)^t M^2 \left(\frac{1}{N}\sum_{j=1}^{N} y_j\right) \\ &+ \sum_{j=1}^{N}\left(y_j - \frac{1}{N}\sum_{k=1}^{N} y_k\right)^t M^2 \left(y_j - \frac{1}{N}\sum_{k=1}^{N} y_k\right) \end{aligned} \tag{4.6}$$

The last term in the above expression is the variation with respect to the average value of the response.[18]

$$\sum_{j=1}^{N}\left(y_j-\frac{1}{N}\sum_{k=1}^{N}y_k\right)^{t}M^2\left(y_j-\frac{1}{N}\sum_{k=1}^{N}y_k\right)$$
$$=\sum_{n=1}^{N}\sum_{i=1}^{r}C_i\left(y_n-\frac{1}{N}\sum_{k=1}^{N}y_k\right)_i^2=(N-1)\sum_{i=1}^{r}C_i s_i^{\,2} \qquad (4.7)$$

The quality loss is written as the sum of three terms: the production cost, the cost associated with the average response that is not the expected one, and the cost arising from the variations with respect to the average value. In other words:

$$L_N=NC_o+N\left(\frac{1}{N}\sum_{j=1}^{N}y_j\right)^{t}M^2\left(\frac{1}{N}\sum_{j=1}^{N}y_j\right)+(N-1)\sum_{i=1}^{r}C_i s_i^{\,2} \qquad (4.8)$$

The average quality loss for the production is the quality loss per unit produced:

$$\frac{L_N}{N}-C_o+\left(\frac{1}{N}\sum_{j=1}^{N}y_j\right)^{t}M^2\left(\frac{1}{N}\sum_{j=1}^{N}y_j\right)+\frac{N-1}{N}\sum_{i=1}^{r}C_i s_i^{\,2} \qquad (4.9)$$

When the production size tends to infinity, the standard deviation for each functional requirement, σ_i, appears in the last term:

$$\frac{L_N}{N}=C_o+\left(\frac{1}{N}\sum_{j=1}^{N}y_j\right)^{t}M^2\left(\frac{1}{N}\sum_{j=1}^{N}y_j\right)+\sum_{i=1}^{r}M_{ii}^{\,2}\sigma_i^{\,2} \qquad (4.10)$$

4.2.3 Process capability index

The process capability index is defined as the quotient between the length of the acceptance interval and six times the standard deviation obtained as a result of the design and production processes (Taguchi, Elsayed and Hsiang, 1989).

$$Cp_i=\frac{\overline{l}_i-\underline{l}_i}{6\sqrt{\dfrac{\sum_{j=1}^{N}\left[(l_i)_j-\frac{1}{N}\sum_{j=1}^{N}(l_i)_j\right]^2}{N-1}}} \qquad (4.11)$$

The response should be made dimensionless by using the variables defined in (3.13).

$$Cp_i = \frac{1}{3\sqrt{\dfrac{\sum_{j=1}^{N}\left[(y_i)_j - \dfrac{1}{N}\sum_{j=1}^{N}(y_i)_j\right]^2}{N-1}}} = \frac{1}{3s_i} \tag{4.12}$$

When the process capability index is higher than 1.0, the process is capable. When the process capability index is equal to 1.0, there is a 0.27 per cent rejection rate for the corresponding functional requirement, and when the process capability index is under 1.0, the process is not capable. In 1980, Japan established a process capability index of 1.33 as a general quality standard. Pioneering companies sought process capacity values of around 2.0. If design activities are involved from the beginning (in order to reduce quality loss), process capacities can reach values as high as 5.0.[19] Note that, according to (4.13), reducing quality loss requires increasing process capacity.

$$\frac{L_N}{N} = C_o + \left(\frac{1}{N}\sum_{j=1}^{N} y_j\right)^t M^2 \left(\frac{1}{N}\sum_{j=1}^{N} y_j\right) + \frac{N-1}{N}\sum_{i=1}^{r}\frac{C_i}{9Cp_i^2} \tag{4.13}$$

A process capability index of around 5.0 means that the value of σ_i is roughly 1/15. In other words, the amplitude of the variation is around 15 times lower than the maximum permitted value. This means going from zero defects to zero variation.

4.2.4 Effect of variation on transfer functions

During design, the product characteristics are given in expression (3.11). During operation, however, the mathematical model shown in (3.11) must be corrected to consider modeling errors (for example, truncation errors) and uncertainties in the knowledge of particular variables (such as noise or inaccuracies). For this reason, the actual characteristics of a product will differ from the calculated ones. To take these effects into account, we will add the noise sources associated with model δy_o and those associated with operation and design parameters δx to transfer function (3.11):[20]

$$y = y_o + \delta y_o + A(x + \delta x) \tag{4.14}$$

In an N-sized production, the response for product number j is:

$$y_j = y_o + \delta_j y_o + A(x + \delta_j x) \tag{4.15}$$

The quality loss associated with the entire production is:[21]

$$\begin{aligned} L_N &= \sum_{j=1}^{N} L(y_j) = NC_o + \sum_{j=1}^{N} y_j{}^t M^2 y_j \\ &= NC_o \\ &\quad + N\left[y_o + Ax\right]^t M^2 \left[y_o + Ax\right] \\ &\quad + \sum_{j=1}^{N} \delta_j y_o{}^t M^2 \delta_j y_o \\ &\quad + \sum_{j=1}^{N} \delta_j x^t A^t M^2 A \delta_j x \\ &\quad + 2\sum_{j=1}^{N} \delta_j y_o{}^t M^2 A \delta_j x \\ &\quad + 2\left[y_o + Ax\right]^t M^2 \sum_{j=1}^{N} \delta_j y_o{}^t \\ &\quad + 2\left[y_o + Ax\right]^t M^2 A \sum_{j=1}^{N} \delta_j x \end{aligned} \tag{4.16}$$

The first addend is the cost of the entire production. The second addend is the cost arising because the response obtained during design did not fall in the center of the intervals formed by the acceptance limits. The third term is the cost arising because the response during operation moves out of the center of the acceptance interval due to noise and errors not considered in the equations in the theoretical model. The fourth term is the cost arising because the response during operation moves out of the center of the acceptance interval due to noise and errors not considered in the operation and design parameters. While the above terms are always positive, the last three can be positive or negative. If $\sum_{j=1}^{N} \delta_j x \neq 0$ occurs in a design, there is a systematic error that can be easily corrected by displacing the intervals where the design parameters vary. In other words, by properly displacing the values $\overline{m}_i$ and $\underline{m}_i$, the average values of the noise affecting the design parameters can be nulled. If these errors are also statistically independent from those in the model, we have:

$$\sum_{j=1}^{N} \delta_j x = 0 \tag{4.17}$$

$$L_N = NC_o + N[y_o + Ax]^t M^2 [y_o + Ax] + \sum_{j=1}^{N} \delta_j y_o{}^t M^2 \delta_j y_o + 2[y_o + Ax]^t M^2 \sum_{j=1}^{N} \delta_j y_o{}^t + \sum_{j=1}^{N} \delta_j x^t A^t M^2 A \delta_j x \tag{4.18}$$

The third term can be developed to show the variation and displacement of the average value.

$$L_N = NC_o + N[y_o + Ax]^t M^2 [y_o + Ax] + N\left(\frac{1}{N}\sum_{k=1}^{N} \delta_k y_o{}^t\right)^t M^2 \left(\frac{1}{N}\sum_{k=1}^{N} \delta_k y_o{}^t\right) + \sum_{j=1}^{N}\left(\delta_j y_o - \frac{1}{N}\sum_{k=1}^{N} \delta_k y_o{}^t\right)^t M^2 \left(\delta_j y_o - \frac{1}{N}\sum_{k=1}^{N} \delta_k y_o{}^t\right) + 2[y_o + Ax]^t M^2 \sum_{j=1}^{N} \delta_j y_o{}^t + \sum_{j=1}^{N} \delta_j x^t A^t M^2 A \delta_j x \tag{4.19}$$

If the values of the operation and design parameters are fixed by the expression $[y_o + Ax]^t = -\sum_{j=1}^{N} \delta_j y_o{}^t / N$, three of the addends in the previous expression would be canceled. Thus, the reduction in quality loss due to the deviation of the average value in the model requires retouching certain design parameters that we will call *adjustment parameters* (or control factors),[22] so that $y_o + Ax = -\sum_{j=1}^{N} \delta_j y_o / N$ is verified. This adjustment can be performed unit by unit. The adjustment parameters can be used to eliminate the offset from the response, but not the variation of the response because there is usually noise that is not present when the unit is being calibrated. Also, noise is usually dynamic or fluctuating, and therefore cannot be eliminated through a simple static calibration operation. The resulting quality loss would be:

$$
\begin{aligned}
L_N &= NC_o \\
&+ \sum_{j=1}^{N}\left(\delta_j y_o - \frac{1}{N}\sum_{k=1}^{N}\delta_k y_o{}^t\right)^t M^2 \left(\delta_j y_o - \frac{1}{N}\sum_{k=1}^{N}\delta_k y_o{}^t\right) \\
&+ \sum_{j=1}^{N}\delta_j x^t A^t M^2 A \delta_j x
\end{aligned} \tag{4.20}
$$

The three terms that remain in the quality loss function are the production cost, the cost associated with the inaccuracy of the transfer functions comprising the model, and the cost associated with the variability of the operation and design parameters. These three terms are always positive, and cannot be nulled. The objective is to minimize them. In particular, correctly choosing the design matrix during the design phase can greatly reduce the last two addends.

4.2.5 Selection of the design matrix

The third addend in expression (4.20) depends on the design matrix and the cost matrix. This expression, which is shown in matrix form, can be written as a product of scalar numbers:

$$
\begin{aligned}
\sum_{j=1}^{N}\delta_j x^t A^t M^2 A \delta_j x &= \sum_{n=1}^{N}\sum_{i=1}^{q}\sum_{j=1}^{r}\sum_{k=1}^{q}\delta_n x_i A_{ji} M_{jj}{}^2 A_{jk}\,\delta_n x_k \\
&= \sum_{i=1}^{q}\sum_{j=1}^{r}\sum_{k=1}^{q} A_{ji} M_{jj}{}^2 A_{jk} \sum_{n=1}^{N}\delta_n x_i \delta_n x_k
\end{aligned} \tag{4.21}
$$

When the noise can be approached by statistically independent probability distributions with a null average (remember that the average was nulled by displacing the intervals), the last sum provides an estimation of the noise standard deviations:[23]

$$
\sum_{n=1}^{N}\delta_n x_i \delta_n x_k = (N-1)s'_{ii}{}^2\delta_{ik} \tag{4.22}
$$

This expression calculates the last term of the quality loss function as:

$$
\begin{aligned}
\sum_{j=1}^{N}\delta_j x^t A^t M^2 A \delta_j x &= \sum_{i=1}^{q}\sum_{j=1}^{r}\sum_{k=1}^{q} A_{ji} M_{jj}{}^2 A_{jk}\,(N-1)s'_{ii}{}^2\delta_{ik} \\
&= (N-1)\sum_{i=1}^{q}\sum_{j=1}^{r} A_{ji}{}^2 M_{jj}{}^2 s'_{ii}{}^2
\end{aligned} \tag{4.23}
$$

The resulting quality loss is:

$$L_N = NC_o + \sum_{j=1}^{N}\left(\delta_j y_o - \frac{1}{N}\sum_{k=1}^{N}\delta_k y_o{}^t\right)^t M^2\left(\delta_j y_o - \frac{1}{N}\sum_{k=1}^{N}\delta_k y_o{}^t\right) + (N-1)\sum_{i=1}^{q}\sum_{j=1}^{r} A_{ji}{}^2 M_{jj}{}^2 s'_{ii}{}^2 \quad (4.24)$$

This expression shows that the quality loss grows along with the production size, cost, noise, and the absolute value of the elements in the design matrix. Due to the construction of the variation intervals for the design parameters (see the definition of the tolerances in Section 2.9), the noise cannot exceed the value 1.0. For this reason, $s'_{ii} \le 1$ is verified. On the other hand, it is normal for the cost arising from non-compliance with a functional requirement to be less than product cost $C_i \le C_o$. Thus, an upper bound for the last term in the quality loss function is:

$$(N-1)\sum_{i=1}^{q}\sum_{j=1}^{r} A_{ji}{}^2 M_{jj}{}^2 s'_{ii}{}^2 \le (N-1)C_o\sum_{i=1}^{q}\sum_{j=1}^{r} A_{ji}{}^2 = (N-1)C_o Tr(AA^t) \quad (4.25)$$

The double sum is the trace of matrix AA^t obtained by multiplying the design matrix by its transpose.[24] Therefore, the larger the trace of the design matrix, the worse the design matrix will be. In general, the quality loss is obtained by performing a trace of matrix $S'^{t1}A^tM^2AS'$, where matrix S' is a diagonal matrix whose elements are the standard deviations of each design parameter.

$$(N-1)\sum_{i=1}^{q}\sum_{j=1}^{r} A_{ji}{}^2 M_{jj}{}^2 s'_{ii}{}^2 = (N-1)\sum_{i=1}^{q}\sum_{j=1}^{r} s'_{ii} A_{ji} M_{jj} M_{jj} A_{ji} s'_{ii} = (N-1)Tr\left[(MAS')^t MAS'\right] \quad (4.26)$$

The matrix MAS' is obtained by scaling the design matrix by rows with the square root of the cost arising from non-compliance with a particular functional requirement, and by columns with the different intensities of the noise present in the design parameters. Finally, the quality loss function can be written as:

$$L_N = NC_o + \sum_{j=1}^{N}\left(\delta_j y_o - \frac{1}{N}\sum_{k=1}^{N}\delta_k y_o{}^t\right)^t M^2\left(\delta_j y_o - \frac{1}{N}\sum_{k=1}^{N}\delta_k y_o{}^t\right) + (N-1)Tr\left[MAS'(MAS')^t\right] \quad (4.27)$$

The above expressions indicate that the trace can be a good indicator for measuring how good or bad a particular design matrix is.[25] Thus, in addition to the indicators I_1 and I_2 defined in (3.50) and (3.51), the following indicator can be used:

$$I_3 = \frac{1}{r(q-1)} \sum_{i=1}^{r} \left(\frac{\sum_{j=1}^{q} A_{ij}^{\ 2}}{\max_j (A_{ij}^{\ 2})} - 1 \right) \tag{4.28}$$

For any matrix, $I_3 \in [-1/(q-1), 1]$ is verified, where $I_3 = 0$ is the optimum value of the indicator. This indicator measures the excess or lack of dominant elements in the design matrix. If the indicator is greater than zero, then $r(q-1)I_3$ is a lower bound for the number of elements that should be eliminated to reach the optimum solution. If the indicator is less than zero, it provides an idea of the number of elements missing in the matrix. As we can see, it is very similar to the indicator I_2 defined in (3.51), and can be used as a substitute for it.[26]

4.3 System design and parameter design

To design is to choose those transfer functions (system design) and parameter values (parameter design) that minimize the quality loss function. Choosing the transfer functions is part of the conceptual design of the system, while choosing the operation and design parameters is part of the detailed design. The detailed design also includes the tolerance design that will be studied in the next section.

The response for specimen j in the functional requirement i is:

$$(y_j)_i = (y_o)_i + (\delta_j y_o)_i + \sum_{k=1}^{q} A_{ik} \left(x_k + (\delta_j x)_k \right) \tag{4.29}$$

Using expressions (3.8) and (3.12), the above equation can be written as:

$$(y_j)_i = \frac{2 f_i(m_o) - \overline{l}_i - \underline{l}_i}{\overline{l}_i - \underline{l}_i} + (\delta_j y_o)_i + \sum_{k=1}^{q} \frac{2}{\overline{l}_i - \underline{l}_i} \left. \frac{\partial f_i}{\partial m_k} \right|_{m_o} \left(m_k - (m_o)_k + \delta_j m_k \right) \tag{4.30}$$

In general, the metric has a cost structure that depends on the design point chosen, and can therefore be written as:

$$L_N = \sum_{j=1}^{N} L(y_j) = NC_o + \sum_{j=1}^{N} y_j^{\ t} M^2 y_j = NC_o(m_o) + \sum_{j=1}^{N} \sum_{i=1}^{r} (y_j)_i^2 C_i(m_o) \tag{4.31}$$

The metric represented by equation (4.31) requires two types of information: information from the customer and information from the design studio and, in general, the entire organization. The first type of information, associated with the customer, consists of the acceptance limits $\overline{l}_i$ and $\underline{l}_i$ shown in (4.30). The second type of information includes the knowledge of cost and the product performance that will be achieved. The performance values are calculated by equation (4.30), which in turn depends on the design matrix, the values adopted for the design parameters, and the noise. These values depend strongly on the system design, and must be estimated or calculated by the engineer or designer. In particular, the conceptual system design sets the number of operation and design parameters, as well as the functions that relate them to the response and cost. It is the engineer's job to know these functions.

$$C_i = C_i(m_o) \quad i = 0,1,\ldots,r \tag{4.32}$$

$$f_i = f_i(m_o) \tag{4.33}$$

$$A_{ij} = A_{ij}(m_o) \tag{4.34}$$

If we take $m = m_o$, the design equations that enable us to perform the parameter design are:[27]

$$\frac{\partial L_N}{\partial (m_o)_k} = 0 \quad k = 1,2,\ldots,q \tag{4.35}$$

With a fixed cost structure and in the absence of noise, these equations lead to the design equations:

$$f_i(m_o) = \frac{\overline{l}_i + \underline{l}_i}{2} \tag{4.36}$$

Equations (4.36), used to find the values of the operation and design parameters that satisfy the user's needs, represent a very well-known design philosophy: classic system design. Normally, there are more parameters to determine than equations to determine them. This means that 'expert' opinions are often required. Based on their experience, such experts manage to choose the value of certain design parameters. However, if the conceptual system design was performed properly, the very conception of the design procedure for maximum quality always has the same number of equations as unknown quantities. The reason is that minimizing the quality loss (4.31) is formally equal to making the partial derivatives of the quality loss function equal to zero with respect to each and every one of the unknown design parameters, generating as many equations as unknown quantities in a natural manner (4.35). Thus, Metric Design replaces equations (4.36) with equations (4.35). When non-compliance with one functional requirement is much costlier than non-compliance with the others, and the noise in one design parameter produces much greater variations than the others, Metric Design will seek a point where the cost and the sensitivity of the transfer function are lower. Metric Design therefore combines design for minimum cost, robust design and classic design. The optimum point for society depends on the cost structure and noise structure.

The fact that a product that perfectly satisfies the customer's needs has a non-null quality loss is immediately deduced from the fact that an implicit customer need is always to have the need satisfied at zero cost.[28] The failure to achieve this is a quality loss in itself. If we also assume that variability always exists, the cost assumed by society must be higher than the product price itself. This last statement is true because the resources consumed are the same that would have been consumed if the product were perfect, but with the aggravating circumstance that it is not. In other words, the quality loss increases as the gap between the product's performance and the user's needs grows. If the gap increases enough, the customer's level of insatisfaction will cause him to reject the product. When rejection occurs, the social cost is twice the cost of the product.[29] The fact that the first product is defective means that another appropriate one must be provided. Who assumes the cost is indifferent in this global conception; what is important is that society had to assume it, which means that it is not free. In the long term, society itself will be affected (the environment, the company, the customer, or all of the above, which is what usually occurs). Defective products, i.e. those falling outside the margin, are not a quality loss problem, but a cost problem (Taguchi, Elsayed and Hsiang, 1989). If non-compliance with specifications requires replacing the entire product with a new one, and the price of the product is independent from the design point, the metric is simplified:

$$\frac{L_N}{NC_o} - 1 = \frac{1}{N}\sum_{j=1}^{N}\sum_{i=1}^{r}(y_j)_i^2 \qquad (4.37)$$

The product cost depends on factors internal to the company, such as the design of the entire organization, which includes the design of the product itself, and external factors such as raw material prices. Choosing design parameters for minimum cost might lead to a design point with maximum variability. The sum of both items is what must be kept to a minimum. For the same reason, a robust design according to (4.37) can produce values farther from those specified by the customer according to equations

(4.36). Normally, this problem is solved if a series of adjustment parameters are chosen during the system design phase that do not affect the cost structure or the noise structure.

The structure of equation (4.31), as a sum of terms that are all positive, recovers the same information obtained in Axiomatic Design. That is, for a design to have the lowest quality loss, the following must be verified: 1) The number of functional requirements must be kept to a minimum (equivalent to Corollary 2). 2) The number of design parameters must be equal to the number of requirements (equivalent to Theorem 4).[30] 3) The intensity of all noise in the adjustment parameters must be low (equivalent to Corollary 7). However, the new information provided is that the cost must be low. This directly affects the tolerance design.

Theorem: If the number of design parameters is lower than the number of functional requirements, a quality loss is incurred.

Proof: This will be proven for two functional requirements, y_1 and y_2 whose values must be $m_{y_1} \pm \Delta_{y_1}$ and $m_{y_2} \pm \Delta_{y_2}$ respectively. The quality loss function is $L = C_1((y_1 - m_{y_1})/\Delta_{y_1})^2 + C_2((y_2 - m_{y_2})/\Delta_{y_2})^2$. If there is a single design parameter, x, related to the functional requirements by the following engineering equations, $y_1 = A_1 x$ and $y_2 = A_2 x$, this can be rewritten as $L = ((x - x_1)/\Delta_1)^2 + ((x - x_2)/\Delta_2)^2$, where $x_1 = m_{y_1}/A_1$, $x_2 = m_{y_2}/A_2$, $\Delta_1 = \Delta_{y_1}/(A_1\sqrt{C_1})$, and $\Delta_2 = \Delta_{y_2}/(A_2\sqrt{C_2})$. The design with the minimum quality loss is the one that verifies $\partial L/\partial x = 0$. The design point $x_0 = (x_1\Delta_2^2 + x_2\Delta_1^2)/(\Delta_1^2 + \Delta_2^2)$ verifies that condition. At that point, the quality loss is $L(x_0) = L_0 = (x_1 - x_2)^2/(\Delta_1^2 + \Delta_2^2)$. Consequently, the quality loss function can be rewritten as $L(x) = L_0 + ((x - x_0)/\Delta_{12})^2$, where $\Delta_{12} = \Delta_1\Delta_2/\sqrt{\Delta_1^2 + \Delta_2^2}$. The first addend is a constant term that represents the economic loss produced by trying to cover two functional requirements with a single design parameter. This quality loss is due to the poor choice of system. It cannot be minimized during the parameter design phase. The second addend measures the quality loss associated with noise and variability, and can be minimized during parameter and tolerance design.

4.4 Tolerance design

It is clear that the tolerances of the product's characteristics significantly affect quality loss. The optimum tolerances are those that minimize total quality loss. The three terms that appear in expression (4.8) are very similar to the three that appear in (4.24). The difference between the two expressions is that the former specifies the quality loss in terms of functional requirement values, while the latter specifies it in terms of design parameters. However, from a formal perspective, the design parameters can be viewed as a set of new functional requirements. We simply have to establish a new metric matrix, M', whose elements are the costs associated with the actions needed to reset the design parameters to their acceptance values. Thus, expression (4.24) could also have been written as:

$$L_N = NC_o + \sum_{j=1}^{N}\left(\delta_j y_o - \frac{1}{N}\sum_{k=1}^{N}\delta_k y_o{}^t\right)^t M^2 \left(\delta_j y_o - \frac{1}{N}\sum_{k=1}^{N}\delta_k y_o{}^t\right) + (N-1)\sum_{i=1}^{q} M'_{ii}{}^2 s'_{ii}{}^2 \tag{4.38}$$

By comparing (4.38) to (4.24), we can find the expression that defines the cost arising from non-compliance with the design parameters:

$$M'_{ii}{}^2 = \sum_{j=1}^{r} A_{ji}{}^2 M_{jj}{}^2 \tag{4.39}$$

If we remember that the elements in the design matrix are given in expression (3.8), equation (4.39) is equal to:

$$C'_i = \sum_{j=1}^{r} \frac{(\overline{m}_i - \underline{m}_i)^2}{(\overline{l}_j - \underline{l}_j)^2}\left(\frac{\partial f_j(m_1,\ldots,m_q)}{\partial m_i}\right)^2 C_j \tag{4.40}$$

It is interesting to isolate the tolerance value affecting the design parameter:

$$\frac{\overline{m}_i - \underline{m}_i}{2} = \frac{\sqrt{C'_i}}{\sqrt{\sum_{j=1}^{r}\left(\frac{2}{\overline{l}_j - \underline{l}_j}\right)^2\left(\frac{\partial f_j(m_1,\ldots,m_q)}{\partial m_i}\right)^2 C_j}} \tag{4.41}$$

Tolerancing Theorem: Strict tolerances should be set for the components that are cheapest to adjust.

Proof: Expression (4.41) shows that the tolerance of a design parameter is proportional to the cost of non-compliance. Thus, the lower the cost of non-compliance, the stricter the tolerance should be.

This is undoubtedly one of the most interesting and powerful results of Metric Design. Because it requires a cost structure, it cannot be derived in the pure context of Axiomatic Design. When there is a single functional requirement, expression (4.41) becomes:

$$\frac{\overline{m}_i - \underline{m}_i}{2} = \frac{\overline{l} - \underline{l}}{2} \frac{1}{\left|\frac{\partial f(m_1,\ldots,m_q)}{\partial m_i}\right|} \sqrt{\frac{C'_i}{C}} \tag{4.42}$$

Expression (4.42) indicates that tolerance selection depends on the sensitivity (as in classic tolerance design) and the cost structure. Expression (4.41) also takes into account the influence of the design parameter on the rest of the functional requirements. By comparing (4.8) and (4.24), we find the relationship between the variations in the functional requirements and design parameters:

$$s_i^{\,2} = \sum_{j=1}^{q} A_{ij}^{\,2} s'^{\,2}_{jj} \tag{4.43}$$

Again, by replacing the element in the design matrix with the value specified in (3.8), we obtain:

$$\left(\frac{\overline{l}_i - \underline{l}_i}{2}\right)^2 s_i^{\,2} = \sum_{j=1}^{q} \left(\frac{\overline{m}_j - \underline{m}_j}{2}\right)^2 \left(\frac{\partial f_i(m_1,\ldots,m_q)}{\partial m_j}\right)^2 s'^{\,2}_{jj} \tag{4.44}$$

Finally, (4.41) enables us to write:

$$\left(\frac{\overline{l}_i - \underline{l}_i}{2}\right)^2 s_i^{\,2} = \sum_{j=1}^{q} \frac{\left(\frac{\partial f_i(m_1,\ldots,m_q)}{\partial m_j}\right)^2 C'_j s'^{\,2}_{jj}}{\sum_{k=1}^{r} \left(\frac{2}{\overline{l}_k - \underline{l}_k}\right)^2 \left(\frac{\partial f_k(m_1,\ldots,m_q)}{\partial m_j}\right)^2 C_k} \tag{4.45}$$

For a single functional requirement, the above equation is reduced to:

$$s^2 = \sum_{j=1}^{q} \frac{C'_j}{C} s'^{\,2}_{jj} \tag{4.46}$$

If all of the design parameters are distributed according to a uniform distribution in the interval defined by their tolerance, $s'_{jj} = 1/\sqrt{3}$, and the above expression:

$$s^2 = \frac{1}{3} \sum_{j=1}^{q} \frac{C'_j}{C} \tag{4.47}$$

When $\sum_{j=1}^{q} C'_j > C$ is met, Metric Design specifies that the response must exceed the acceptance limits. In other words, because the operation to calibrate the final product is cheaper than the operation to calibrate each component, it is preferable to manufacture the components with less precision and incorporate a final calibration operation. On the contrary, when $\sum_{j=1}^{q} C'_j < C$, calibrating the whole is more expensive than calibrating each of the components. In this situation, the tolerance of each component must be made as strict as possible (while ensuring that the production costs do not increase the quality loss).

4.5 Robust design

Robust designs are those in which the response does not vary, even in the presence of noise (Wu and Wu, 2000). When the noise is considerable and the transfer functions are not linear, Metric Design tends to reduce the elements in the design matrix that produce the greatest variation in the response. Artificially increasing the noise is a very useful technique for quickly locating the points where the design is robust.

The functional requirements are given by expression (3.13), and are related to the design parameters, defined according to (3.12), through the transfer functions. Thus, quality loss function (4.3) uses the design parameters as arguments:

$$L(m_1,\ldots,m_q) = C_o + \sum_{i=1}^{r} C_i \left[\frac{f_i(m_1,\ldots,m_q) - \dfrac{\overline{l}_i + \underline{l}_i}{2}}{\dfrac{\overline{l}_i - \underline{l}_i}{2}} \right]^2 \qquad (4.48)$$

The average quality loss value for the entire production, when the production size tends to infinity, is determined by the probability density function of the design parameters according to the expression:

$$\frac{L_N}{N} = \frac{\sum_{j=1}^{N} L_j}{N} = \int_{-\infty}^{\infty} dm_1 \ldots \int_{-\infty}^{\infty} dm_q \, pdf(m_1,\ldots,m_q) L(m_1,\ldots,m_q) \qquad (4.49)$$

A robust design is insensitive to noise, and therefore not very dependent on the particular form of the probability density function. As an initial approach, we will assume that the noise affecting the design parameters is independent and uniformly distributed. Under these conditions, the multiple integral that appears in the above equation is transformed into:

$$\frac{L_N}{N} = \frac{\int_{-\delta_1}^{\delta_1} dm_1 \ldots \int_{-\delta_q}^{\delta_q} dm_q\, L(m_1,\ldots,m_q)}{\prod_{i=1}^{q} 2\delta_i} \tag{4.50}$$

If L is a continuous function of its arguments, the average value theorem establishes that there is a point in the integration volume where it is verified that the value of the integral coincides with the value of the function multiplied by the volume. In order to boost the noise, we will estimate that value using the average taken by the function on the boundary generated by the maximum variation. The response of functional requirement i depends on the noise present in the transfer function arguments according to (3.12) and (3.13).

$$y_i = \frac{f_i\left((m_o)_1 + \delta_1 x_1,\ldots,(m_o)_q + \delta_q x_q\right) - \frac{\overline{l}_i + \underline{l}_i}{2}}{\frac{\overline{l}_i - \underline{l}_i}{2}} \tag{4.51}$$

where $\delta_i = \frac{\overline{m}_i - \underline{m}_i}{2}$ is the maximum variation of design parameter m_i. Thus, the terms $\delta_1 x_1$ to $\delta_q x_q$ can be interpreted as the noise affecting the average values of the design parameters. The maximum variation of design parameter m_i is obtained when $x_i = \pm 1$. In other words, we will approach (4.50) by:[31]

$$\frac{L_N}{N} = \frac{1}{2^q}\sum_{i=0}^{2^q-1} L\left((m_o)_1 + (-1)^{bit_0(i)}\delta_1,\ldots,(m_o)_q + (-1)^{bit_{q-1}(i)}\delta_q\right) \tag{4.52}$$

In this expression, the function $bit_i(j)$ returns the value of the bit located in the ith position in the binary expression for natural number j. If we introduce (4.48) into the above expression:

$$\frac{L_N}{N} = C_o + \frac{1}{2^q}\sum_{j=0}^{2^q-1}\sum_{i=1}^{r} C_i \left[\frac{f_i\left((m_o)_1 + (-1)^{bit_0(j)}\delta_1,\ldots,(m_o)_q + (-1)^{bit_{q-1}(j)}\delta_q\right) - \frac{\overline{l}_i + \underline{l}_i}{2}}{\frac{\overline{l}_i - \underline{l}_i}{2}}\right]^2 \tag{4.53}$$

This form of the quality loss function is very useful for designing. The values of δ_i can be obtained from expressions (4.41) or set as arbitrarily high noise in order to force a robust design. Once the amplitude of the noise has been set, we can minimize (4.53) in order to find the optimum values for the design parameters. Obviously, the physical limitations or constraints affecting the design parameters are conditions that must be checked during the optimization process. As we can see in (4.53), each quality loss function calculation requires evaluating each component in the transfer function 2^q times. The design of experiments makes it possible to reduce the number of evaluation points in the function, and therefore the computational costs if the number of design parameters is very high (see Appendix).

4.6 Cost-effectiveness of advanced design techniques

The previous section shows that Metric Design constitutes an optimization problem (non-linear and subject to a large number of constraints). Approaching and solving it can quickly overwhelm the computational capacity and drastically increase the number of man-hours that must be devoted to design. This additional design cost must initially be paid, while the benefits derived from a Metric Design begin when production is in the customer's hands. This leads to a key question: How do we determine whether a product should be designed using Axiomatic Design, Metric Design or, in general, an Advanced Design technique? In other words, when is it cost-effective to assume the higher design costs associated with Advanced Design techniques in exchange for future benefits? The answer to this question depends strongly on which product and industrial sector are considered; however, we will use Metric Design to make a first attempt.

We will now consider the following problem. We need to design a design process for a product whose response must remain within the acceptance intervals set by the customer. In this problem, there are several needs. On the one hand, there are needs associated with the external customer, who will be the product user. On the other hand, there is an internal customer, who will be in charge of designing the product, and must do so in the shortest time possible.

Let l be the product's response and $[\underline{l}, \overline{l}]$ the acceptance interval. For convenience, we will define $l_o = (\overline{l} + \underline{l})/2$ and $\Delta = (\overline{l} - \underline{l})/2$. On the other hand, we will assume that if the product's response falls outside the acceptance interval, loss C_P is incurred, equal to the product cost. In an N-sized production, product number j, which reaches the customer, incurs a quality loss given by $L_j = C_P + C_P\,((l_j - l_o)/\Delta)^2$. If production is centered, the total loss associated with the entire production is:

$$L_N = \sum_{j=1}^{N} L_j = NC_P\left(1+\frac{\sigma^2}{\Delta^2}\right) \tag{4.54}$$

where σ is the standard deviation of that functional requirement induced by varying the rest of the parameters.

Let t be the design time and $[0, \bar{t}]$ the acceptance interval.[32] Suppose that if the design time exceeds the upper limit, an economic loss is incurred that is equal to the cost of the design process, C_D. The quality loss incurred by the design process is:

$$L_D = C_D + C_D\left(\frac{t-0}{\bar{t}}\right)^2 \tag{4.55}$$

The total quality loss for the design process and the production will be:

$$L_T = L_N + L_D = NC_P + C_D + NC_P\left(\frac{\sigma}{\Delta}\right)^2 + C_D\left(\frac{t}{\bar{t}}\right)^2 \tag{4.56}$$

Consider two designs for the same product, one using Advanced Design (AD) and the other Classic Design (CD) techniques. The product is characterized by the functional requirements, and the design process by the time required for the design. On the other hand, the market where that product will be released is the same in both design situations. Thus, the number N of units sold in both situations will be the same if product price C_P is also the same in both situations. The customer does not vary, and the rejection margin for functional requirement Δ is also the same in both cases. The same is true of rejection margin $\bar{t}$ associated with the design time, which reflects the needs of the members of the organization. In other words, the time that can be devoted to the design without missing the chance to be released on the market is also the same. However, the standard deviation of the response of the product designed using a classic design technique will be higher than that of a product designed using an advanced design technique. Let σ_{CD} be the standard deviation of a classic design, and $\sigma_{AD} < \sigma_{CD}$ the standard deviation of an advanced design. Suppose that the design time for an advanced design process is greater, $t_{AD} > t_{CD}$. Similarly, suppose that the cost associated with an advanced design is also higher, $C_{AD} > C_{CD}$. The quality loss in both cases will be:

$$L_{CD} = NC_P + C_{CD}\left(\frac{t_{CD}}{\bar{t}}\right)^2 + NC_P\left(\frac{\sigma_{CD}}{\Delta}\right)^2 \tag{4.57}$$

$$L_{AD} = NC_P + C_{AD}\left(\frac{t_{AD}}{\bar{t}}\right)^2 + NC_P\left(\frac{\sigma_{AD}}{\Delta}\right)^2 \tag{4.58}$$

For the advanced design to be cost-effective, $L_{AD} < L_{CD}$ must be met, in other words:

$$C_{AD}\frac{t_{AD}^2}{\bar{t}^2} - C_{CD}\frac{t_{CD}^2}{\bar{t}^2} < NC_P\frac{\sigma_{CD}^2 - \sigma_{AD}^2}{\Delta^2} \tag{4.59}$$

Compliance with this inequation requires:

1. Very costly products compared to the cost required to design them, $C_P \gg C_{AD} > C_{CD}$.
2. Products for which large series are expected to be made, $N \gg 1$.
3. Markets that will change very little over time, $\bar{t} \to \infty$.
4 Very strict acceptance limits by the customer, $\Delta \to 0$.

One example of this type of product could be precision instrumentation. Indeed, the cost of the equipment can be much higher than that of the design, due to the quality of the elements comprising it and the cost of the calibration and certification tests. The series produced are large due to the universality of the measuring instruments, measuring needs do not change drastically, and high precision is required, $\sigma_{CD} \gg \Delta \gg \sigma_{AD}$. At the other extreme, it is not worth devoting much effort to the design of very cheap equipment for a small number of undemanding customers in rapidly changing markets. Neither is it cost-effective for products where the design cost is in the order of, or much higher than, the product cost, the production size is low, and a classic design is capable of ensuring precision, $\sigma_{CD} \ll \Delta$. One example of this type of product could be a satellite. In these situations, the optimum design would be a classic design of the complete system, and later an advanced design of those subsystems that will be employed systematically. Obviously, these results are only an initial approach to the problem of designing the design process. In special situations, the metric can provide different results depending on where the actual boundary for each of the parameters intervening in the problem is located.

4.7 Example application

The best design process is the one that enables us to meet our stated objectives.[33] To evaluate the effectiveness of an advanced design process, it is useful to consider the metaphor of an archer trying to hit a target.[34] Before each shot, the archer might take small steps forward or backward to modify his longitudinal position, change the inclination of the bow, and tense or loosen the string on the bow. In other words, he can change the distance to the target and the exit angle and velocity of the arrow. The archer must assemble the bow, aim, and shoot. The question is how to optimize these processes in order to hit the target the highest number of times. To answer this, we will use Robust Design equation (4.53), derived from Metric Design.

As an academic example,[35] suppose that the target size is 4 cm, and that it is located at a height of 2 m over the bow and a minimum longitudinal distance of 1 m from the bow. (The maximum distance is not specified, but is assumed to be less than 4 m.) The maximum exit velocity from the bow is 10 m/s. The archer who hits the target the highest number of times, out of a total of 12, wins the championship. Suppose that every time the archer aims the bow for the first time, he makes mistakes in the order of a few centimeters in the longitudinal position, several degrees in the inclination, and several meters per second in the exit velocity. The dimensionless transfer function is:

$$y(x,\alpha,V)=\frac{\left(\tan\alpha-\dfrac{gx}{2V^2\cos^2\alpha}\right)x-\dfrac{\overline{l}+\underline{l}}{2}}{\dfrac{\overline{l}-\underline{l}}{2}} \tag{4.60}$$

From the above, we know that the upper acceptance limit is $\overline{l}$ = 2020 mm, the lower acceptance limit is $\underline{l}$ = 1980 mm, the longitudinal position is $x \in$ (1000 mm, 4000 mm) with an error of δ_x = 20 mm, the angle is $\alpha \in$ (0,90°) with an error of δ_α = 2°, and the exit velocity is $V \in$ (0,10 m/s) with an error of δ_v = 2 m/s. The expected quality loss function according to (4.53) is:

$$\frac{\dfrac{L_N}{N}-C_o}{C_1}=\frac{1}{8}\sum_{j=0}^{7}\left[y\left(x+(-1)^{bit_0(j)}\delta_x,\alpha+(-1)^{bit_1(j)}\delta_\alpha,V+(-1)^{bit_2(j)}\delta_V\right)\right]^2 \tag{4.61}$$

In this function, C_0 is the cost associated with participating in the championship, and C_1 is the cost associated with not winning the championship. Inaccuracies in the position, inclination and velocity cause the archer to miss one of the 12 shots required to win the championship. The costs associated with these design parameters are therefore $C_x = C_\alpha = C_v = C_1/12$. The optimum tolerances are obtained from (4.42).

$$\frac{\overline{x}-\underline{x}}{2}=\frac{1}{\sqrt{12}\left|\dfrac{\partial y}{\partial x}\right|},\ \frac{\overline{\alpha}-\underline{\alpha}}{2}=\frac{1}{\sqrt{12}\left|\dfrac{\partial y}{\partial \alpha}\right|},\ \frac{\overline{V}-\underline{V}}{2}=\frac{1}{\sqrt{12}\left|\dfrac{\partial y}{\partial V}\right|} \tag{4.62}$$

The corrections that cause a shot on the acceptance limit to hit the center of the target are:

$$\Delta_x = \frac{1}{\left|\frac{\partial y}{\partial x}\right|}, \Delta_x = \frac{1}{\left|\frac{\partial y}{\partial \alpha}\right|}, \Delta_V = \frac{1}{\left|\frac{\partial y}{\partial V}\right|} \tag{4.63}$$

The result obtained by minimizing the quality loss function (4.61) is shown in Table 4.4.

According to this table, the archer must perfect his technique to achieve position errors under 5.9 mm and angle errors under 0.083 degrees. However, he need not worry about exit velocity errors, as long as the velocity is high enough.[36] The optimum strategy for an inexperienced archer is to set the position around 2026 mm and aim. It is usually easier to achieve a precision in the order of half a centimeter than in one-tenth of a degree. Then, if he misses the shot, he should correct his longitudinal position and try not to modify the angle. If he does not perfect his technique and the errors are in the order of the initial ones, the quality loss is roughly $L_N / N = C_o + 49.4C_1$. This value is much higher than the cost of not winning the championship, indicating that the archer will lose the championship much more often than he will win it. If he were able to perfect his technique and reduce his errors to those indicated by the optimum tolerances, the new strategy would be to move about one meter farther away from the target (see Table 4.5). The quality loss would be $L_N / N = C_o + 0.14C_1$, indicating a higher number of victories than defeats in the championships.

Table 4.4 **Optimum parameter values for an inexperienced archer**

	Value	Noise	Minimum	Maximum	Correction	Tolerance
x (mm)	2026	20	1000	4000	20.4	5.9
α (degrees)	44.4	2	5	85	0.289	0.083
V (m/s)	10	2	0	10	253000	73100

Table 4.5 **Optimum parameter values for a trained archer**

	Value	Noise	Minimum	Maximum	Correction	Tolerance
x (mm)	3013	5.89	1000	4000	30.1	8.7
α (degrees)	33.6	0.0833	5	85	0.264	0.076
V (m/s)	10	2	0	10	156000	45000

Table 4.6 **Optimum parameter values for an archer without angular positioning errors**

	Value	Noise	Minimum	Maximum	Correction	Tolerance
x (mm)	4000	5.89	1000	4000	40.0	11.5
α (degrees)	26.6	0	5	85	0.229	0.066
V (m/s)	10	2	0	10	102000	29400

Finally, by completely canceling the error in the angle, but not in the longitudinal position, the strategy shown in Table 4.6 requires increasing the shooting distance as much as possible.[37] The quality loss would be $L_N / N = C_o + 0.02C_1$.

4.8 Notes

1. Kar (2000) has linked Axiomatic Design and the Taguchi method through information content.
2. Fowlkes and Creveling (1995) explains that there are several business behaviors that frequently lead to a variation in components and performance producing a uniform probability distribution between the upper and lower specification limit. For example: 1. Companies maximize productivity by placing everything produced within the specified limits on the market without considering differences in quality. 2. They use manufacturing strategies that take into account the known deviations in the processes, and maximize the output within the specified limits. 3. Components that do not meet the specifications are remade so that they comply. All of these behaviors lead to a customer being just as likely to purchase a product on one of the acceptance limits as one in the center of the acceptance interval.
3. Universality inevitably requires converting all quality indicators to the same unit so that they are universally understandable and universally comparable.
4. The acceptance limits (the customer's tolerances) can usually be defined as the level of performance at which 50 per cent of the customers are unsatisfied.
5. This definition clarifies that each specific product within an N-sized production will be operating under certain service conditions and subject to particular sources of noise, and will constitute an economic loss, due to its purchase and use, that society must assume. This is adapted from Taguchi, Elsayed and Hsiang (1989: 2) 'Quality loss is defined as the loss a product costs society from the time the product is released for shipment.'
6. Obviously, although the definition is universal, the specific value of the loss will depend on the customer, product and situation.
7. Metric Design appreciates the difference in quality between two designs, both with a null information content, if one is located near the extreme of the acceptance interval and the other near the center of the acceptance interval. This edge effect is reflected in the quality loss.

8. This postulate implicitly assumes that the acceptance intervals are symmetrical with respect to the optimum point. Taguchi, Elsayed and Hsiang (1989) and Fowlkes and Creveling (1995) introduce the quality loss function for non-symmetrical cases.
9. This definition of quality replaces other, less universal, definitions. For example, it is possible to define quality as the set of characteristics that enable us to distinguish an object as better or worse than others. However, this type of definition says little about what the optimum set of characteristics is (the one with a higher or lower number of elements), how to compare those elements (whether they should be compared one by one, globally, etc.), or how to establish the comparison pattern. Logothetis (1992) presents a broad perspective of the concept of quality.
10. These inevitable variations are produced by factors out of our control, which we will refer to as noise.
11. Variation is the chronic disease of industry. It is dangerous because if accepted as such, it institutionalizes waste and recovery. In the long term, all of this creates quality loss (Taguchi, Elsayed and Hsiang, 1989). Metric Design seeks to immunize products and processes in order to minimize this (because eliminating it is virtually impossible).
12. An experiment is a test in which some input variables of a process or system are modified so that we can observe and identify changes in the process or system's response. The design of experiments is the tool needed to obtain the desired results with the smallest number of experiments possible, and to minimize the variance in the coefficients obtained by regression (see Appendix). A reduction in the number of experiments acts directly on the time and effort employed, resulting in reduced costs. Minimizing the variance acts directly on the quality. In this way, the design of experiments is a very powerful tool for increasing the value of a product. The objective is to find a new point of operation and design with less quality loss.
13. When the response is measured using the variables given in (2.32) or (3.13), the Metric Design Postulate states that the lowest quality loss is obtained when vector $y^t = (y_1, \ldots, y_r)$ is null. Remember that these variables are equal to zero if the response is in the center of the acceptance interval, 1.0 if they are in the upper extreme of the acceptance interval, and –1.0 if they are in the lower extreme of the acceptance interval. These values are important because they make it possible to replace generic function (4.2) with (4.3).
14. The engineering metrics employed in robust product design are derived from this equation (Taguchi, Elsayed and Hsiang, 1989; Taguchi and Wu, 1991; Taguchi, Konishi and Wu, 1992; Fowlkes and Creveling, 1995). The quality loss function widely used in the literature is such that $L(0) = 0$, and assumes that a perfect product has no quality loss even if its cost is exorbitant. This simplification is only correct for studying the effect of variability on quality loss, and is a simplified form of the original function. When cost is one of the objectives to be met in the product design or organization, equation (4.3) must be used, where $L(0) \neq 0$. The result of mixing the Mahalanobis' distance and Taguchi's ideas is described in Taguchi, Chowdhury and Wu (2001) and Taguchi and Jugulum (2002).

15. The variables used to perform the quality loss calculation are usually not deterministic. For example, the acceptance limits or the costs deriving from a rejection may vary from one customer to another. For this reason, such values are usually defined as those that a significant number of customers (for example, 50 per cent) would accept. Thus, the acceptance limits are those values for which a significant number of customers would take an economic action due to a deviation from the product's expected performance. The value of this economic action is each of the C_i constants. On the other hand, in certain situations, the quality loss function might not be symmetrical. There can be several reasons for this. For example, if the diameter of a shaft is much larger than the nominal value, only the verification and later machining costs are lost. However, if the diameter is much smaller than the nominal value, the entire part might be lost. In these situations, the cost can be approached by the average value of the upper and lower costs associated with rejection.
16. Because the company is striving to satisfy the customer, product performance will come very close to the desired values, so $y \simeq 0$. In this situation, the quality loss function can be expanded in a power series around $y = 0$. The theorem assumes that the quality loss function can be expanded for all of its arguments within the acceptance interval. This type of argument is often called 'nominal-the-best' because the optimum value for the functional requirement has higher and lower bounds. Dimensions, temperatures, flows, etc., are examples of this type. However, some quality characteristics are non-negative, and are considered better as the values get lower. In other words, the target or nominal value is zero. These are called 'smaller-the-best' characteristics. Impurities, noise, weight, volume and dilation are some examples of this type. (Reliability and efficiency have a target value equal to 1.0, and are not defined over that. By taking one minus reliability and one minus efficiency, they are transformed into 'smaller-the-best' characteristics.) Strictly speaking, because they are not defined for non-negative values, the Metric Design Postulate does not eliminate the linear term; for these, expression (4.3) could contain a linear term. By convention, the quality loss that they produce is usually defined as proportional to y_i^2. The last type is 'larger-the-better,' for example, mechanical resistance, lifetime, etc. The optimum nominal value would be infinite. By convention, the quality loss that they produce is usually considered proportional to $1/y_i^2$. Benavides (2004) shows how the quality loss produced by a fatigue lifetime is proportional to t^{-s}, where t is the lifetime and s is the damage exponent associated with a Weibull distribution (see Appendix).
17. Because matrix B is symmetrical, there is an orthonormal matrix P that verifies $B = PDP^t$, where D is a diagonal matrix formed by the eigenvalues of B. If all of its eigenvalues were equal to λ, we would have $D = \lambda I$ where I is the identity matrix; consequently, $B = PDP^t = \lambda PP^t = \lambda I$ would be diagonal. Therefore, if B is not diagonal, it must have at least two different eigenvalues. In addition, if B is not diagonal, then P cannot be diagonal either (in another case, $B = PDP^t = DPP^t = D$ would be diagonal because diagonal matrices commute). As long as P is not diagonal, the requirements are coupled because it would be necessary to modify several at once to find the optimum solution.

18. For random variable y, $s=\sqrt{\frac{1}{N-1}\sum_{j=1}^{N}\left(y_j-\sum_{k=1}^{N}y_k/N\right)^2}\xrightarrow[N\to\infty]{}\sigma$ is a random variable whose average value is the standard deviation of y (see Appendix).
19. A uniform distribution whose limits are the acceptance limits –1 and +1 has a standard deviation of $s_i=1/\sqrt{3}$, i.e., a process capability index equal to Cp_i = 0.577. To obtain Cp_i = 2 (Cp_i = 5), the deviation must be $s_i=0.29/\sqrt{3}$ ($s_i=0.12/\sqrt{3}$). In all three cases, the percentage of rejections is null. If the distributions were normal rather than uniform, the percentage of rejections would be 8.3 per cent, $2.0\cdot10^{-7}$ per cent and $7.3\cdot10^{-49}$ per cent for Cp_i = 0.577, 2 and 5, respectively.
20. The response and the operation and design parameters are expressed as dimensionless variables calculated using expressions (3.12) and (3.13). This means that all statistical variables deriving from them (such as the standard deviation) will also be dimensionless.
21. We do not consider the non-linear terms in the transfer function (see (4.14) and (4.15)). This means that we also do not consider the effects of these terms on the quality loss function (4.16). Chapter 6 will show that the non-linear terms lead to a coupling between the mean and the standard deviation.
22. The variation of these adjustment parameters must follow the adjustment directions calculated in Section 3.8.
23. We use s' for the standard deviation (of a sample) of the design parameters, and s for the standard deviation (of a sample) of the functional requirements (Equation (4.8)).
24. $Tr(A^tA)=\sum_{i=1}^{q}\sum_{j=1}^{r}A_{ji}A_{ji}=\sum_{j=1}^{r}\sum_{i=1}^{q}A_{ji}A_{ji}=Tr(AA^t)$
25. Metric Design modifies design matrix A given in (3.8) to give MAS', incorporating the noise intensities by columns. By rows, it incorporates the square root of the cost arising from non-compliance with the functional requirement.
26. The difference between I_2 and I_3 is that the absolute value of each element has been replaced by its square. This tends to reduce the influence of non-dominant elements when calculating I_3. In Chapter 3, we assumed $I_2 \in [0,1]$ rather than $I_2 \in [-1/(q-1),1]$ because the number of design parameters cannot be lower than the number of functional requirements.
27. Benavides (2006) shows the application of these equations to the selection of bearings with minimum mass, maximum lifetime (under a superficial fatigue failure mode), and different cost structures.
28. The minimum price is a customer need and a need imposed by the minimization of quality loss. Thus, an objective (explicit or implicit) is to obtain the product at zero cost. If the price of the product is C_o, it is reasonable to expect that the customer will reject the product if it is supplied at a higher price. The rejection margin is therefore the product price itself, and the economic loss (if any) is the product price.
29. It was necessary to create two identical products in order for only one of them to work. This is an initial approach to the problem, which assumes that

rejection of a product constitutes its complete loss. The general definition is given by equation (4.3), where, in general, $C_i \neq C_o$.

30. The number of design parameters must be kept to a minimum in order to avoid variation. Nonetheless, unlike Axiomatic Design, Metric Design can use the excess design parameters to find a design point with a lower quality loss. Therefore, the higher the number of design parameters, the greater the flexibility, and the greater the flexibility, the greater the possibility of finding a better design point. Nevertheless, Axiomatic Design specifies that the number of adjustment parameters and functional requirements must be the same.
31. This method for choosing points is taken from the design of experiments (see Appendix). The points on the boundary are distributed so that: 1) the associated experimental matrix is orthogonal, and 2) all of the design parameters provide information at each point. These conditions keep the experimental error to a minimum.
32. Design time is a 'smaller-the-best' type requirement. In other words, the lower limit of the acceptance interval coincides with the value sought.
33. The product designed will be considered good if it satisfies the needs, functional requirements and constraints in every instance where it is used.
34. Fowlkes and Creveling (1995) use this metaphor with a catapult.
35. These dimensions make it possible to perform the experiment in a classroom with a small bow that shoots pieces of chalk. The students are divided into groups that compete against each other. The experiment is conducted on two different days. The first phase takes place before studying advanced design techniques, and the second after learning them. The students choose a strategy, establish a bow configuration, and make the shots.
36. The exit velocity is ensured using a set of n rubber bands. Each rubber band has a natural length l_o and rigidity k. They are anchored to form a triangle with base $2H$ and height V. The bow tenses, causing the height of the triangle to become $V + L$ without changing the value of the base. The force created by the set of rubber bands is $F(L) = 2nk[2\sqrt{H^2+(V+L)^2} - l_o](V+L)/\sqrt{H^2+(V+L)^2}$. The accumulated elastic energy is

$$E = 2nk\left[H\left(\left(\frac{V+L}{H}\right)^2 - \left(\frac{V}{H}\right)^2\right) - l_o\left(\sqrt{1+\left(\frac{V+L}{H}\right)^2} - \sqrt{1+\left(\frac{V}{H}\right)^2}\right)\right]$$

If the mass of the projectile is m and that of the thruster is M, the exit velocity is $V = \sqrt{2E/(m+M)}$. When $L \ll V$, we have $E = 2nk\frac{VL}{H}\left[2 - \frac{l_o}{H}\Big/\sqrt{1+\left(\frac{V}{H}\right)^2}\right]$.

37. It is an interesting exercise to verify that when the noise in the longitudinal position is null and the noise in the angular position is not, the best shooting point is the one given by $\alpha = 45°$.

5

Reliability-based design

Abstract: This chapter addresses the approximate calculation of the probability that a functional requirement will exceed a particular acceptance limit. The chapter discusses such calculations in a general way in order to present a set of ideas that are also valid for calculating the probability of failure associated with a certain failure mode for a component.

Key words: success, failure, probability, reliability model.

5.1 Objective of reliability calculations

In Chapter 3, Axiomatic Design was discussed as a general design tool. Axiom 2 in particular, regarding information content, leads to the design of systems with a minimum number of functional requirements (Corollary 2), and sets the broadest acceptance intervals possible (Corollary 6). Chapter 4, on the other hand, introduces Metric Design for minimizing the quality loss in designs. Part of the quality loss comes from uncontrolled noise. Tolerance design is a technique commonly used to reduce quality loss during design. Specifying a stricter tolerance for a design parameter means reducing the variation of that parameter. However, as discussed in Chapter 3, the design parameters on one level of the design hierarchy become functional requirements on the next level of the hierarchy. Establishing strict tolerances for the design parameters therefore contradicts Corollary 6 (Axiom 2). Consequently, this practice should be avoided during design.

Combining both design needs, or 1) meeting the customer's specifications the highest number of times possible, and 2) doing so with the best design, requires assuming, on the one hand, that tolerances are necessary, and on the other, that they should be eliminated. This situation shows that there is an optimum value for the number of tolerances and their amplitude. Without yet defining a cost structure, we can initially approach tolerance design based on reliability criteria. Reliability calculations also establish the percentage of 'failures' associated with a particular design. If this number is too high, the quality loss will also be high, Axiom 2 will not be verified, and the design should be discarded.

The objective of this chapter is to provide the basic tools for estimating the order of magnitude of the expected reliability for a design.

5.2 Definition of reliability

We intuitively understand the reliability of a system as the probability that it will 'survive.'[1] In a broad sense, reliability is the probability that it will satisfy the customer's needs. For this reason, reliability is closely related to the Information Axiom and expression (3.28), which calculates when the system's response falls within the acceptance intervals. Chapter 4 also includes, as a quality loss, the effect of noise on non-compliance with specifications. However, noise is not the only factor leading to insatisfaction. For example, a sudden unexpected breakage of the product (a catastrophic failure) will also cause the response to fall outside of the acceptance interval. A system's failure mode is understood to be the mechanism leading to the failure to satisfy a need.

5.2.1 Failure mode

Failure mode refers to each of the different events leading to a response outside of the acceptance intervals. In complex engineering systems, there are different failure modes that can affect one or more needs. Consequently, the number of possible failure modes is quite a bit higher than the number of needs met. This means that before calculating overall system reliability, we have to calculate the reliability associated with each failure mode. Because there is usually one failure mode with a much lower reliability than the others, this can be considered the *critical failure mode*. The reliability associated with the critical failure mode, if there is only one, will be very close to the component reliability.

Thus, a preliminary analysis of the component must be performed during design to determine the possible failure modes. This process is called failure mode analysis (FMA). Each failure mode will have an associated 'failure' or 'breakage' criterion from the corresponding branch of engineering. For example, exceeding a certain Von Mises stress, exceeding a particular service temperature, exceeding a maximum admissible deformation, etc. The contribution of failure mode analysis with respect to Axiomatic Design and Metric Design lies in accepting that there are system behaviors not considered in the system transfer function. In other words, the transfer function employed to design the system in Metric Design or Axiomatic Design might no longer be valid during operation due to a change produced during such operation. If this change, which occurred during system operation, is large enough to drastically alter the component's response and cause it to fall outside of the acceptance intervals, this is a failure mode that must be studied.[2]

In general, failure mode i will have a failure criterion that can be expressed as follows: there is a failure if $g_i(x) < 0$.[3] g_i is a function that depends on a different number of k_i variables for each failure mode:

$$g_i\colon \mathbb{R}^{k_i} \rightarrow \mathbb{R},\ x \mapsto g_i(x) \tag{5.1}$$

Finding the probability of failure (one minus the reliability) associated with each failure mode means calculating the probability:

$$\Pr[g_i\,(x) < 0] = F_i = 1 - R_i \tag{5.2}$$

The solution's reliability is found by combining all of the probabilities of failure associated with the failure modes identified during the failure mode analysis process (the number of failure modes will be called *FMA*). If the different failure modes are independent, the reliability is:[4]

$$R = \prod_{i=1}^{FMA} R_i = \prod_{i=1}^{FMA} (1 - F_i) = 1 - \sum_{i=1}^{FMA} F_i + \sum_{i=1}^{FMA} \sum_{j=i+1}^{FMA} F_i F_j + \ldots \tag{5.3}$$

Figure 5.1 shows the relationship between the different failure modes and the reliability of a design.

From all of the above, we can deduce the need to establish relatively simple calculation methods that enable us to calculate $\Pr[g_i(x) < 0]$. The information obtained through such calculations is:

- Determination of the critical failure mode (with the modes in increasing order of criticalness).

Figure 5.1 Reliability calculation process for a solution affected by different independent failure modes

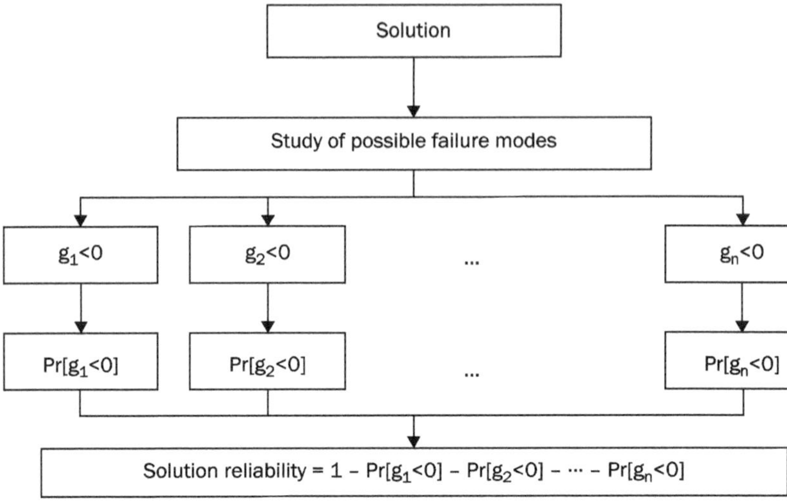

- Order of magnitude of the reliability in those modes.
- Sensitivity to noise of the reliability of each failure mode for each operation and design parameter.
- Number of tolerances needed and their order of magnitude.

5.2.2 *Effect of operation and design parameter variability*

A particular case is a product failure due to the noise present in the operation and design parameters. Such noise is partially responsible for the product's quality loss, and must be controlled by an adequate tolerance design as described in Chapter 4. However, even if the product's quality loss is minimal, non-compliance with specifications might be an event with a non-null probability. The methods described in the following sections can be applied to such a case by simply replacing the general failure criterion $g_i < 0$ with $y_i + 1 < 0$ or $y_i - 1 > 0$, depending on whether functional requirement y_i (defined by (3.13)) fails to reach the lower acceptance limit, or exceeds the upper acceptance limit.

5.3 Calculating the probability of failure

The problem involves solving the following multiple integral:

$$F_i = \Pr\left[g_i(x) < 0\right] = \int\limits_{g_i(x)<0} dx_1 \ldots dx_{k_i} pdf_i(x) \tag{5.4}$$

In the above integral, pdf_i is the probability density function associated with the operation and design parameter vector appearing in failure mode $g_i(x) < 0$. It is a weighting that considers that there may be regions of space that are inaccessible or lack physical meaning (in these regions, the distribution function would be null or practically null), and more likely regions where the designer has set the range of variation for vector x (vector of operation and design parameters). If design parameter x_i is defined according to expression (3.12), the distribution function would need to have a maximum value when m_i was near $(\overline{m}_i + \underline{m}_i)/2$, and would need to be very low at extremes $\underline{m}_i$ and $\overline{m}_i$ of the acceptance interval. If we define the function:

$$\phi : \mathbb{R} \rightarrow \mathbb{R}, a \mapsto \phi(a) = \begin{cases} 1 \;\; if \;\; a < 0 \\ 0 \;\; otherwise \end{cases} \tag{5.5}$$

The probability of failure is therefore the mean value of function $\phi[g_i(x)]$.

$$F_i = \int\limits_{g_i(x)<0} dx_1 \ldots dx_{k_i} pdf_i(x) = \int\limits_{\mathbb{R}^{k_i}} dx_1 \ldots dx_{k_i} \phi[g_i(x)] pdf_i(x) = E\left[\phi[g_i(x)]\right] \tag{5.6}$$

The variance of function $\phi[g_i(x)]$ is:

$$\begin{aligned} &\int\limits_{\mathbb{R}^{k_i}} dx_1 \ldots dx_{k_i} \left[\phi[g_i(x)] - E\left[\phi[g_i(x)]\right]\right]^2 pdf_i(x) \\ &= \int\limits_{\mathbb{R}^{k_i}} dx_1 \ldots dx_{k_i} \phi[g_i(x)] pdf_i(x) - E\left[\phi[g_i(x)]\right]^2 = F_i(1 - F_i) \end{aligned} \tag{5.7}$$

Expression (5.6) approaches the probability of failure through an N-sized sample obtained from population x. If $\xi_j \in \mathbb{R}^{k_i}$, where $j = 1,. . .,k_i$, are the points in the sample, an estimation of the probability of failure is:

$$\hat{F}_i = \frac{1}{N} \sum_{i=1}^{N} \phi[g_i(\xi_i)] \tag{5.8}$$

According to (5.6) and (5.7), the random variable $\hat{F}_i$ belongs to a population with the mean F_i and standard deviation $\sqrt{F_i(1-F_i)/N}$. An estimation of the variance of $\phi[g_i(x)]$ is:

$$\frac{1}{N-1} \sum_{i=1}^{N} \left(\phi[g_i(\xi_i)] - \hat{F}_i\right)^2 = \frac{1}{N-1} \sum_{i=1}^{N} \left(\phi[g_i(\xi_i)] - \hat{F}_i^2\right) = \frac{N}{N-1} \hat{F}_i(1 - \hat{F}_i) \tag{5.9}$$

From the above, we can deduce that the variance of $\hat{F}_i$ is $\hat{F}_i(1 - \hat{F}_i)/(N - 1)$. Consequently, the following random variable has a null mean and unit standard deviation:

$$z = \sqrt{N-1}\frac{\hat{F}_i - F_i}{\sqrt{\hat{F}_i(1-\hat{F}_i)}} = \sqrt{N-1}\sqrt{\frac{\hat{F}_i}{1-\hat{F}_i}}\left(1-\frac{F_i}{\hat{F}_i}\right) \tag{5.10}$$

For a confidence level of $1 - \alpha$, the estimation error is in the order of:

$$\left|\frac{F_i}{\hat{F}_i} - 1\right| = \frac{1}{\sqrt{N-1}}\sqrt{\frac{1-\hat{F}_i}{\hat{F}_i}}\left|z_{\alpha/2}\right| \tag{5.11}$$

The random variable $\hat{F}_i$ is constructed as the average of a set of N identical and independent variables that can take the values one or zero. Thus, if the number of points used in the average is very high, the Central Limit Theorem (see Appendix) establishes that the distribution of random variable z is a normal $N(0,1)$. If the confidence level is set to 96 per cent ($\alpha = 0.04$ and $z_{\alpha/2} \simeq 2$), the error would be:[5]

$$\left|\frac{F_i}{\hat{F}_i} - 1\right| \simeq \frac{2}{\sqrt{N}}\sqrt{\frac{1-\hat{F}_i}{\hat{F}_i}} \tag{5.12}$$

When the reliability is very high, an estimation of the number of points required in the sampling of x is:

$$N \simeq \frac{4}{\left(\frac{F_i}{\hat{F}_i} - 1\right)^2 \hat{F}_i} \tag{5.13}$$

Therefore, good estimations of very high reliabilities require a large number of points in the sample. Estimating a failure probability in the order of 10^{-3} with an error in the order of 10 per cent requires evaluating the function $\phi[g_i(x)]$ more than 400,000 times.[6] In most cases, such a high number of evaluations makes it impossible to calculate the integral through direct simulation. For this reason, approximate calculation methods like the one described below are enormously useful.

5.4 First-Order Reliability Model (FORM)

This section addresses the calculation of integral (5.4) using an approximate, first-order technique. The result is a method capable of estimating the reliability and its sensitivities with respect to the operation and design parameters. This enables us to select the critical failure mode and the nominal values and tolerances for the parameters.

5.4.1 Linear failure criterion with two variables

Take a linear failure criterion with only two variables, A and B, where the first defines the value obtained, and the second establishes the minimum acceptable value. Thus, if the value obtained exceeds the minimum acceptable value, the configuration is safe.

$$g = A - B < 0 \Rightarrow \text{Failure!} \tag{5.14}$$

Both are random variables whose mean and standard deviation are given by η_A and σ_A for A, and η_B and σ_B for B. To solve the problem, it is advisable to enter dimensionless variables:

$$A' = \frac{A-\eta_A}{\sigma_A} \quad B' = \frac{B-\eta_B}{\sigma_B} \tag{5.15}$$

With these new variables, the boundary between the region where the failure occurs and the safe zone is given by:

$$g = \sigma_A A' + \eta_A - \sigma_B B' - \eta_B = 0 \Rightarrow B' = \frac{\sigma_A}{\sigma_B} A' + \frac{\eta_A - \eta_B}{\sigma_B} \tag{5.16}$$

As we can see, given the linearity of the failure criterion, the boundary is a straight line. Because the addition of two random variables is another random variable whose mean is the sum of the means and whose variance is the sum of the variances, we can define the new random variable g' with a null mean and unit standard deviation:

$$g' - \frac{g-(\eta_A - \eta_B)}{\sqrt{\sigma_A^2 + \sigma_B^2}} \tag{5.17}$$

The failure criterion expressed in this new variable is:

$$g = g'\sqrt{\sigma_A^2 + \sigma_B^2} + \eta_A - \eta_B < 0 \Rightarrow g' < -\frac{\eta_A - \eta_B}{\sqrt{\sigma_A^2 + \sigma_B^2}} \tag{5.18}$$

The distance, in variables A' and B', from the origin of the coordinates to the failure boundary is the value that appears in (5.18) with the opposite sign. For this reason, this distance is called the *reliability*

Figure 5.2 Diagram showing the boundary surface with the most probable point (MPP) on it

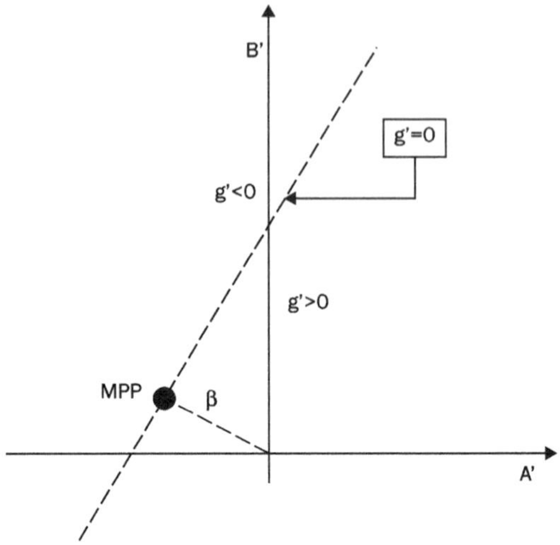

indicator. If we call this distance β, as shown in Figure 5.2, we have:

$$\beta = \frac{\eta_A - \eta_B}{\sqrt{\sigma_A^2 + \sigma_B^2}} \tag{5.19}$$

Finally, the probability of failure can be calculated as:

$$F = \Pr[g' < -\beta] \tag{5.20}$$

This calculation requires that the distribution function for random variable g' be known. Because we are in the initial phases of the design process, where the level of information is low, such distributions are rarely known. However, to estimate the orders of magnitude for the reliability, we can assume a normal distribution (it could come from normal distributions in A and B and a linear model). If we consider a normal distribution for g', the probability of failure would be calculated directly from the normal tables:

$$F = \Pr_{N(0,1)}[-\beta] = \frac{1}{\sqrt{\pi}} \int_{-\infty}^{-\beta} e^{-\xi^2}\, d\xi \tag{5.21}$$

5.4.1.1 Safety factor

Expressions (5.21) and (5.19) indicate that one way to decrease the probability of failure is by increasing the value of $\eta_A - \eta_B$ (high reliabilities require $\eta_A - \eta_B \gg \sqrt{\sigma_A^2 + \sigma_B^2} > 0$.) When $\eta_A \gg \eta_B$, the probability of failure tends to zero; when $\eta_A = \eta_B$, the probability of failure is 0.5; and when $\eta_A \ll \eta_B$, the probability of failure tends to one. If we define the safety coefficient as the quotient $FS = \eta_A / \eta_B$ (where $\eta_B > 0$), reliability indicator (5.19) is written as $\beta = (FS-1)\eta_B / \sqrt{\sigma_A^2 + \sigma_B^2}$.[7] Note that a safety factor much higher than one separates the distributions in such a way that the overlap in the density function tails decreases. This is true in most situations because the noise is usually much lower than the averages. For this reason, increasing the value of the safety factors decreases the probability of failure.[8] The designer therefore has two tools at his disposal for increasing reliability: imposing high safety coefficients and imposing strict tolerances that reduce the values of σ_A and σ_B.[9]

5.4.2 Linear failure criterion with multiple variables

In the case of a failure criterion whose boundary is a hyperplane, the failure occurs when:

$$g(x) = \sum_{j=1}^{k} a_j x_j < 0, \quad x^t = (x_1, \ldots, x_k) \in \mathbb{R}^k \tag{5.22}$$

In this expression, $a_1, \ldots, a_k$ are constants. If each of the x_j variables is a random variable distributed by a normal with a mean of η_j and a standard deviation of σ_j, the variable g is a random variable distributed by the normal $N\left[\sum_{j=1}^{k} a_j \eta_j, \sqrt{\sum_{j=1}^{k} a_j^2 \sigma_j^2}\right]$. If the failure mode is rewritten using the variables $x'_j = (x_j - \eta_j)/\sigma_j$, the reliability indicator is the distance from point $x'_1 = \ldots = x'_k = 0$ to the hyperplane. If we call this distance β, equation (5.23), the probability of failure is $F = \Pr_{N(0,1)}[-\beta]$, equation (5.21).

$$\beta = \frac{\sum_{j=1}^{k} a_j \eta_j}{\sqrt{\sum_{j=1}^{k} a_j^2 \sigma_j^2}} \tag{5.23}$$

5.4.3 Non-linear failure criterion with multiple variables

This corresponds to Hasofer-Lind's theory (Cruse, 1997). It is a generalization of the previous case, where the hypersurface separating the safe zone from the rest has the following general form:

$$g(x) < 0, x \in \mathbb{R}^k \tag{5.24}$$

The reduced operation and design parameter vector is defined as:

$$x'^t = \left(\frac{x_1 - \eta_1}{\sigma_1}, \ldots, \frac{x_k - \eta_k}{\sigma_k} \right) \tag{5.25}$$

The procedure involves:

1. Checking that midpoint $\eta^t = (\eta_1, \ldots, \eta_k)$ (in reduced variables, point $x' = 0$) is a safe configuration, i.e. $g(\eta) > 0$.
2. Establishing the failure criterion $g(x) < 0$ in reduced variables as $g'(x') < 0$.
3. Calculating the reduced vector x' that produces the minimum distance from the origin to the hypersurface $g'(x') = 0$. This point is called the most probable point.
4. Calculating the reliability indicator, the minimum distance to the hypersurface:

$$\beta = \sqrt{x'^t x'} \tag{5.26}$$

5. Calculating the probability of failure as $F = \Pr_{N(0,1)}[-\beta]$ given by (5.21).

The algorithm for calculating the most probable point is based on Taylor's development of the failure criterion up to first-order terms (Cruse, 1997). Thus, in iteration $n + 1$, the failure criterion will be:

$$g'(x'^{(n+1)}) = g'(x'^{(n)}) + \nabla g'(x'^{(n)})^t (x'^{(n+1)} - x'^{(n)}) = 0 \tag{5.27}$$

From here, it is possible to obtain:

$$\nabla g'(x'^{(n+1)}) = \nabla g'(x'^{(n)}) \tag{5.28}$$

$$\nabla g'(x'^{(n)})^t x'^{(n+1)} = \nabla g'(x'^{(n)})^t x'^{(n)} - g'(x'^{(n)}) \tag{5.29}$$

On the other hand, the sphere centered on the origin of radius $\beta = \sqrt{x'^t x'}$ and surface $g'(x') = 0$ must have the same tangent plane. We must therefore verify that their gradient vectors are parallel. Hence,

$$x'^{(n+1)} = \frac{\left| x'^{(n+1)} \right|}{\left| \nabla g'(x'^{(n+1)}) \right|} \nabla g'(x'^{(n+1)}) \tag{5.30}$$

Using the above expressions, we can write:

$$x'^{(n+1)} = \frac{\left|x'^{(n+1)}\right|}{\left|\nabla g'(x'^{(n)})\right|} \nabla g'(x'^{(n)}) \tag{5.31}$$

$$\left|\nabla g'(x'^{(n)})\right| \left|x'^{(n+1)}\right| = \nabla g'(x'^{(n)})^t x'^{(n)} - g'(x'^{(n)}) \tag{5.32}$$

Finally, the new point will be (Cruse, 1997; Rackwitz and Flessler, 1978):

$$x'^{(n+1)} = \frac{\nabla g'(x'^{(n)})^t x'^{(n)} - g'(x'^{(n)})}{\left|\nabla g'(x'^{(n)})\right|^2} \nabla g'(x'^{(n)}) \tag{5.33}$$

This equation establishes an iterative procedure that can begin with a starting point, for example, $x'^{(0)} = 0$, and must stop when the variation $\beta^{(n+1)} / \beta^{(n)} - 1$ is lower than a preset value.[10] Expressions (5.21), (5.26) and (5.33) are a powerful design tool that provides the order of magnitude of the reliability for each failure mode. They can therefore be used to classify the different failure modes from the highest to the lowest criticalness.

5.4.3.1 Sensitivity analysis

Equations (5.21), (5.26) and (5.33) give the probability of failure associated with a particular failure mode. The sensitivity of this probability with respect to the different operation and design parameters is:

$$\frac{\partial F}{\partial \eta_i} = \frac{\partial \Pr_{N(0,1)}[-\beta]}{\partial x'_i} \frac{\partial x'_i}{\partial \eta_i} = \frac{e^{-\beta^2}}{\sqrt{\pi}} \frac{\partial \beta}{\partial x'_i} \frac{1}{\sigma_i} = \frac{e^{-\beta^2}}{\sigma_i \sqrt{\pi}} \frac{\partial \sqrt{x'^t x'}}{\partial x'_i} = \frac{e^{-\beta^2}}{\sigma_i \sqrt{\pi}} \frac{x'_i}{\beta} \tag{5.34}$$

These sensitivity factors combine the sensitivity of the function $g'(x')$, the importance of the failure mode, and the standard deviation of each operation and design variable. The dimensionless form of the sensitivity factors is:

$$\frac{\partial F}{\partial \eta_i} \sigma_i = \frac{e^{-\beta^2}}{\sqrt{\pi}} \frac{x'_i}{\beta} \tag{5.35}$$

These values make it possible to write a design matrix where the functional requirements are the probabilities of failure associated with each failure mode, and the design parameters are the mean values of the operation and design variables adimensionalized with the standard deviation.

5.4.3.2 Example application

Suppose that a prismatic part with a cross-section of A is subject to tensile load Q. The failure occurs when the tension generated exceeds elastic limit Y. The failure of the part therefore occurs when $g = AY - Q < 0$ is verified. The three variables are distributed normally, with a mean and standard deviation that enable us to write the following dimensionless variables:

$$A' = \frac{A - \eta_A}{\sigma_A},\ Y' = \frac{Y - \eta_Y}{\sigma_Y} \text{ and } Q' = \frac{Q - \eta_Q}{\sigma_Q} \tag{5.36}$$

The failure criterion for the dimensionless variables is:

$$g' = (\eta_A + \sigma_A A')(\eta_Y + \sigma_Y Y') - (\eta_Q + \sigma_Q Q') \tag{5.37}$$

The gradient vector is:

$$\nabla g' = \begin{Bmatrix} \dfrac{\partial g'}{\partial A'} \\ \dfrac{\partial g'}{\partial Y'} \\ \dfrac{\partial g'}{\partial Q'} \end{Bmatrix} = \begin{Bmatrix} \sigma_A(\eta_Y + \sigma_Y Y') \\ \sigma_Y(\eta_A + \sigma_A A') \\ -\sigma_Q \end{Bmatrix} \tag{5.38}$$

Equation (5.33) applied to this problem is

$$\begin{Bmatrix} A' \\ Y' \\ Q' \end{Bmatrix}^{(n+1)} = J^{(n)} \begin{Bmatrix} \sigma_A(\eta_Y + \sigma_Y Y') \\ \sigma_Y(\eta_A + \sigma_A A') \\ -\sigma_Q \end{Bmatrix}^{(n)} \tag{5.39}$$

where $J^{(n)}$ has the expression

$$J = \frac{\sigma_A \sigma_Y A'Y' - \eta_A \eta_Y + \eta_F}{\sigma_A^2(\eta_Y + \sigma_Y Y')^2 + \sigma_Y^2(\eta_A + \sigma_A A')^2 + \sigma_F^2} \tag{5.40}$$

Table 5.1 shows the solution for the case $\eta_A = 1$, $\sigma_A = 0.05$, $\eta_Y = 40$, $\sigma_Y = 5$, $\eta_Q = 20$, and $\sigma_Q = 4$, which displays a safety factor of $FS = \eta_A \eta_Y / \eta_Q = 2$. After four iterations,[11] the error in the reliability indicator is negligible, and the probability of failure obtained is in the order of 10^{-3}. The sensitivity analysis shows that in order to reduce the probability of failure, A and Y must be increased, and Q must be reduced. However, it is preferable to modify the elastic limit before the area or force. For this reason, the correct selection of the material and the thermal treatment applied to it will have an important influence on reliability.

This failure mode can occur in the impact of a check valve, such as those used in mechanical injection systems on diesel engines (shown in Figure 5.3), where the variability in the tensile load comes from the

Table 5.1 Result of the iterations performed with expression (5.39)

Iteration	A'	Y'	Q'	$\eta_A = 1$, $\sigma_A = 0.05$, $\eta_Y = 40$, $\sigma_Y = 5$, $\eta_Q = 20$, and $\sigma_Q = 4$					
0	0.0000	0.0000	0.0000	J	β	A'/β	Y'/β	Q'/β	F
1	–0.8889	–2.2222	1.7778	–0.4444	2.9814	–0.2981	–0.7454	0.5963	0.00143
2	–0.6387	–2.2779	1.9071	–0.4768	3.0496	–0.2258	–0.7470	0.6254	0.00115
3	–0.6783	–2.2891	1.8966	–0.4741	3.0491	–0.2225	–0.7507	0.6220	0.00115
4	–0.6768	–2.2898	1.8962	–0.4740	3.0491	–0.2220	–0.7510	0.6219	0.00115

Figure 5.3 Schematic of a transfer valve used in a mechanical fuel injection system for diesel engines

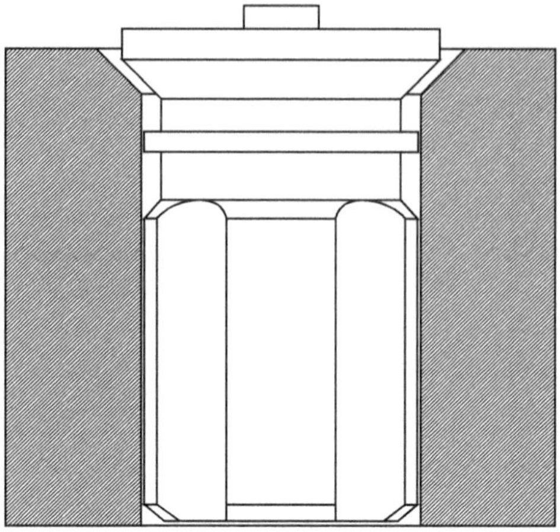

variability in the valve's speed as it strikes its seat.[12] This speed, in turn, depends on the injection pressure, the elastic moduli of the valve and its seat, the valve mass, etc. A piston engine rotating at 3000 rpm with an approximate lifetime of 2000 h produces 180 million strikes on one of these valves (in a four-stroke engine, a strike is produced with every two rotations of the crankshaft). In order for the number of breakages associated with this failure mode to not be significant during that period, the probability of failure for that part must be below 1/180,000,000.

5.5 Semi-empirical reliability model

The theory presented above is a static theory because time is not considered unless the engineering function associated with the failure mode includes it as one of the variables. If time does not appear in the failure criterion, the model does not consider specimen aging. However, the lifetime of a specimen is a random variable with a probability density function that must take into account the effects of aging. Due to their relevance, there are many reliability studies for situations where the load is not completely constant. Some useful references for 'step-stress' type models are Balakrishnan (2009), Balakrishnan, Xie and Kundu (2009), Balakrishnan and Han (2009), Balakrishnan, Zhang and Xie (2009), Balakrishnan et al. (2007), Kateri and Balakrishnan (2008), Han et al. (2006), Nelson (2004) et al. (2004) and Khamis and Higgins (1998).

This Section discusses a reliability model with three constants that must be determined empirically, or by comparison with other similar devices and failure modes. This enables us to reproduce purely dynamic phenomena such as specimen aging with loads that vary over time. The density function used is the Weibull distribution for variable loads (see Appendix). When a specimen is subjected to variable load $Q(L)$, which changes throughout specimen lifetime L, the reliability can be written using the following model involving constants s, p and C:[13]

$$R(L) = e^{\ln(0.9)\, s \int_0^L \left(\frac{Q(\tau)}{C}\right)^{ps} \tau^{s-1} d\tau} \tag{5.41}$$

The designer must know how the three constants in the model, s, p and C, vary depending on the operation and design parameters for a given failure mode. In practice, these models are unknown for most devices. However, the designer can estimate these values based on similar designs or obtain them from the component manufacturer. For some devices and failure modes, these three parameters measure quite different effects (Benavides, 2010). Thus, parameter s, related to aging, basically depends on the material and the treatments it is subjected to; exponent p, related to load concentrations at critical points on the device, depends on the local geometry around the stress concentration points; and constant C, related to how the external load is distributed over the entire device, depends on the design of the entire device. For other devices and failure modes, this distinction is not clear, and the three parameters depend strongly on all of the design parameters.[14] Nonetheless, even if the designer does not know the precise expression for the constants, an estimation is sufficient for determining the influence of the operation and design parameters on reliability. The following example shows how to do this.

5.6 Example application: influence of radial clearance on bearing life with a surface fatigue failure mode

A bearing satisfies the need to keep a point on a spin axis in a predetermined position. Thus, bearing failure occurs when the axle cannot spin freely, or when the position of the point on the axle falls outside of the acceptance limits. The bearing maintains the position of that point on the axle through contact between several elements: the inner raceway, the surface of the rolling elements, and the outer raceway. If the shape of any of these surfaces changes, the point on the axle may leave the region specified by the acceptance limits. Because the element surfaces determine whether the need is met, any event affecting the surface geometry can cause a failure mode. Thus, we can cite corrosion, adherence, abrasion, erosion and surface fatigue as some of the most common failure modes.[15] The designer can minimize or cancel the first four modes listed with an adequate supply of lubricant, and by installing seals and filters to keep out contaminants and corrosive agents. However, for a given load, the contact between the balls and raceways generates a small area of contact that becomes smaller as the elastic modulus of the materials increases. The contact pressure is therefore very high. The maximum shear stress under the surface can be high enough to initiate cracks. Every time the ball passes through the load point, a load cycle is produced that can cause the crack to grow until it reaches the surface, at which point a small amount of material will be dislodged. If the bearing is heavily loaded, this failure mode is difficult to eliminate. The designer must therefore choose operation and design parameters that will ensure the expected lifetime and reliability. The aim of this example is to consider a model that allows the designer to determine the main parameters (design, production, assembly and operation) affecting bearing life, and evaluate their influence.

If we load a bearing with a still outer raceway and an inner raceway rotating at a constant speed, and a radial force with a constant magnitude and direction, then expression (5.41) is reduced to:

$$\left(\frac{\ln R(L)}{\ln 0.9}\right)^{1/s} = L\left(\frac{Q}{C}\right)^{p} \quad (5.42)$$

If we now set the bearing reliability to 0.9, the bearing life can be obtained as follows:

$$L_{10} = \left(\frac{C}{Q}\right)^{p} \quad (5.43)$$

where the bearing life subscript indicates that this lifetime has a probability of failure of 10 per cent. Constant C depends on the entire design, and therefore on the radial clearance of the bearing when the constant was determined. For each radial clearance, there will be one constant. If ε is a dimensionless measure of clearance (see Eq. (5.47)), (5.43) is therefore:

$$L_{10}(\varepsilon) = \left(\frac{C(\varepsilon)}{Q}\right)^{p} \quad (5.44)$$

We will call the diametral clearance x_0 (and the radial clearance $x_0 / 2$) for the bearing mounted on its housing, and x_Q will be the axle displacement due to external load Q imposed on the shaft. Figure 5.4 shows a graphic

Figure 5.4 **Schematic of a bearing showing the raceways (continuous lines) and two rolling elements (shaded elements)**

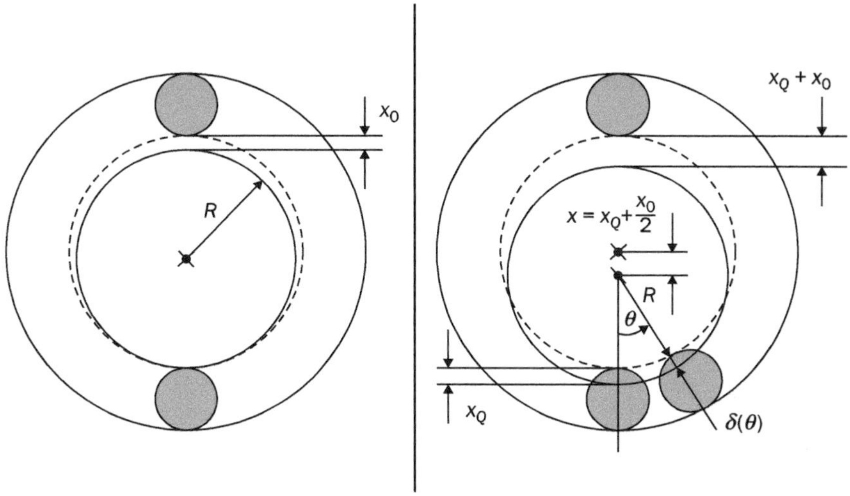

Note: The discontinuous line is an imaginary line generated by moving the rolling elements over the outer raceway. The space created by radially displacing the inner raceway over the rolling elements is the diametral clearance. The figure on the left shows the bearing with no load, while the one on the right shows it loaded.

Source: Adapted from Harris (2001).

representation of the rolling elements and raceways before and after loading the bearing. In the representation of the loaded bearing, only the rolling elements are deformed (the outer and inner rings are infinitely rigid). This hypothesis will enable us to create a model of the influence of radial clearance on bearing life. Figure 5.5 shows a detailed view of the deformed region in accordance with this hypothesis.

According to Figure 5.5, the distribution of the deformation of the rolling elements (as per the cosine theorem, and considering that the offset of both rings is much less than any of their radii) is as follows:

$$\delta(\theta) = x\cos\theta - \frac{x_0}{2} = x(\cos\theta - 1 + 2\varepsilon) \tag{5.45}$$

$$x = x_Q + \frac{x_0}{2} \tag{5.46}$$

$$\varepsilon = \frac{x_Q}{2x} \tag{5.47}$$

When $\varepsilon = 0^+$, the diametral clearance tends to infinity. For $\varepsilon < 0.5$, there is a positive diametral clearance that leaves play between some of the

Figure 5.5 **Main parameters defining the geometry of the area where the rolling elements are compressed**

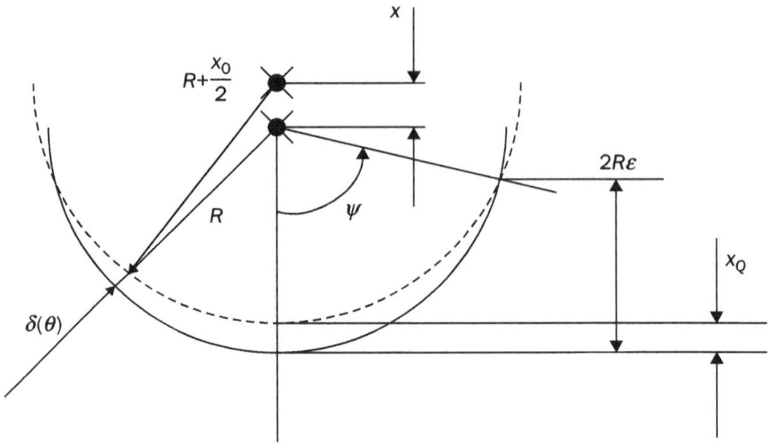

rolling elements and the raceways. When $\varepsilon = 0.5$, the diametral clearance is null, and the deformed portion extends from $-\pi/2$ to $+\pi/2$. When $\varepsilon > 0.5$, the diametral clearance becomes negative, causing a slight interference that increases until $\varepsilon = 1$, at which point the entire perimeter of the raceways experiences interference even though the bearing is loaded radially. If we assume that there are infinite rolling elements, the angular load distribution on the raceway is continuous and different from zero at all points between angles $-\psi$ and $+\psi$ (see Figure 5.5), where ψ is the angle given by the following function:

$$\Psi(\varepsilon) = \begin{cases} \arccos(1-2\varepsilon) & si\ 0 \le \varepsilon < 1 \\ \pi & si\ 1 \le \varepsilon \end{cases} \tag{5.48}$$

Each of the rolling elements trapped in the deformed region displays a point contact transmitting a load proportional to a power of the deformation to which it is subjected (this hypothesis will be justified later). Thus, distributed load $q(\theta)$ is:

$$q(\theta) = K_n\delta^n = K_n x^n(\cos\theta - 1 + 2\varepsilon)^n \tag{5.49}$$

The radial force borne by the bearing will be the one resulting from the vertical component of this distributed load, i.e.:

$$Q = \int_{-\Psi(\varepsilon)}^{\Psi(\varepsilon)} q\cos\theta\, d\theta = K_n x^n \int_{-\Psi(\varepsilon)}^{\Psi(\varepsilon)} (\cos\theta - 1 + 2\varepsilon)^n \cos\theta\, d\theta \tag{5.50}$$

If there are Z rolling elements, the load supported by the most heavily loaded rolling element will be:

$$Q_M(\varepsilon) = K_n x^n (2\varepsilon)^n \frac{2\pi}{Z} \tag{5.51}$$

The two expressions above enable us to eliminate the proportionality constant K_n, so that a relationship appears between the external load applied to the bearing and the load on the most heavily loaded rolling element:

$$Q = \frac{Z}{2\pi} Q_M(\varepsilon) I_1(\varepsilon, n) \tag{5.52}$$

where function $I_1(\varepsilon,n)$ is the following integral:

$$I_1(\varepsilon, n) = \int_{-\Psi(\varepsilon)}^{\Psi(\varepsilon)} \left(\frac{\cos\theta - 1 + 2\varepsilon}{2\varepsilon} \right)^n \cos\theta \, d\theta \tag{5.53}$$

By replacing the load in expression (5.44), we obtain the lifetime as a function of radial clearance:

$$L_{10}(\varepsilon) = \left(\frac{C(\varepsilon)}{\frac{Z}{2\pi} Q_M(\varepsilon) I_1(\varepsilon, n)} \right)^p \tag{5.54}$$

It is plausible to assume that the failure is caused by stress on the most heavily loaded element. That load can then be used in an expression similar to (5.44). For this to occur, the following must be verified:

$$C(\varepsilon) = C^* \frac{Z}{2\pi} I_1(\varepsilon, n) \tag{5.55}$$

Where C^* is a constant that depends on the rest of the design, surface treatments, thermal treatments, and the mechanical properties of the part. The radial clearance modifies the power dissipated by friction, and therefore the temperature of the raceways. This temperature variation on the raceways has two effects on the life of the bearing: 1) it causes thermal expansions that can alter the radial clearance, and 2) it modifies the mechanical properties of the surfaces. The first effect can be calculated by changing the value of ε. However, the second effect must be calculated by changing the value of constant C^*. Experience (Harris, 2001) shows that material hardness is one of the dominant effects on this constant. For this reason, the standards specify a minimum surface hardness of 58 Rockwell C for the tests that determine bearing life. If the hardness, RC, increases or decreases, so will the life of the bearing, according to the following experimentally obtained formula:

$$\frac{C^*(RC(\varepsilon))}{C^*(58)} = \left(\frac{RC(\varepsilon)}{58} \right)^{3.6} \tag{5.56}$$

The hardness of a typical steel used in bearing production decreases as the temperature on the raceway surfaces increases, according to the following experimentally obtained expression (the approximation is valid for 25°C < T < 460°C; T denotes the temperature on the Celsius scale).[16]

$$RC(\varepsilon) = RC(25°C) - 1.46 \cdot 10^{-2}(T(\varepsilon) - 25°C) - 1.32 \cdot 10^{-6}(T(\varepsilon) - 25°C)^2 \quad (5.57)$$

To make comparisons, we will assume a reference test where the radial clearance is null (ε = 1/2) and the raceway temperature is the value provided by a hardness of 58. In this situation, the above expression enables us to write:

$$58 = RC(25°C) - 1.46 \cdot 10^{-2}(T(1/2) - 25°C) - 1.32 \cdot 10^{-6}(T(1/2) - 25°C)^2 \quad (5.58)$$

Based on the two expressions above, and neglecting the non-linear term, we have:

$$RC(\varepsilon) = 58 - 1.46 \cdot 10^{-2}(T(\varepsilon) - T(1/2)) \quad (5.59)$$

Hence, when the radial clearance varies, the constant varies according to:

$$\frac{C^*(RC(\varepsilon))}{C^*(RC(1/2))} = \left(1 - 2.52 \cdot 10^{-4}\left(T(\varepsilon) - T(1/2)\right)\right)^{3.6} \quad (5.60)$$

The raceway temperature depends on the friction because friction torque $T_f(\varepsilon)$ generates heat on the surfaces. If the bearing turns at angular velocity ω, the heat generated is $\omega T_f(\varepsilon)$. This heat is evacuated to the outside by conduction and convection. If $T(\varepsilon)$ is the raceway temperature and T_∞ is the machine temperature, the heat evacuated is proportional to the difference $T(\varepsilon)-T_\infty$. If we add constant α to the model, the raceway temperature can be written as:

$$T(\varepsilon) = T_\infty + \alpha\omega T_f(\varepsilon) \quad (5.61)$$

The friction model that we will use assumes that the friction force is friction coefficient μ_f multiplied by the normal load. The normal load is the load distributed on the rolling elements, shown in equation (5.49). Thus, if r is the mean raceway radius, the friction torque is:

$$T_f(\varepsilon) = 2\pi r\mu_f \int_{-\Psi(\varepsilon)}^{\Psi(\varepsilon)} q\, d\theta = 2\pi r\mu_f K_n x^n \int_{-\Psi(\varepsilon)}^{\Psi(\varepsilon)} (\cos\theta - 1 + 2\varepsilon)^n\, d\theta \quad (5.62)$$

Expression (5.50) makes it possible to eliminate constant K_n (taking into account (5.53)):

$$T_f(\varepsilon) = 2\pi r \mu_f Q \frac{I_2(\varepsilon,n)}{I_1(\varepsilon,n)} \tag{5.63}$$

$$I_2(\varepsilon,n) = \int_{-\Psi(\varepsilon)}^{\Psi(\varepsilon)} \left(\frac{\cos\theta - 1 + 2\varepsilon}{2\varepsilon} \right)^n d\theta \tag{5.64}$$

On the other hand, the test without radial clearance (ε = 1/2) enables us to calculate the constant associated with heat transfer. For this test, (5.65) and (5.63) provide the needed constant:

$$\alpha\omega 2\pi r \mu_f Q \frac{I_2(1/2,n)}{I_1(1/2,n)} = T(1/2) - T_\infty = \Delta T(1/2) \tag{5.65}$$

The model for lifetime verifies (5.52), (5.54), (5.55), (5.60), (5.61), (5.63) and (5.65), which can be summarized in the following three equations:

$$L_{10}(\varepsilon) = \left(\frac{C^*(\varepsilon)}{Q_M(\varepsilon)} \right)^p \tag{5.66}$$

$$Q_M(\varepsilon) = \frac{Q}{I_1(\varepsilon,n)} \frac{2\pi}{Z} \tag{5.67}$$

$$C^*(\varepsilon) = C^*(1/2) \left[1 - 2.52 \cdot 10^{-4} \Delta T(1/2) \left(\frac{I_1(1/2,n) I_2(\varepsilon,n)}{I_1(\varepsilon,n) I_2(1/2,n)} - 1 \right) \right]^{3.6} \tag{5.68}$$

The lifetime given by (5.66) can be adimensionalized using its reference value (ε = 1/2):

$$\frac{L_{10}(\varepsilon)}{L_{10}(1/2)} = \left(\frac{Q_M(1/2)}{Q_M(\varepsilon)} \frac{C^*(\varepsilon)}{C^*(1/2)} \right)^p \tag{5.69}$$

This model (equations (5.67) to (5.69)) predicts that the higher the value of $\Delta T(1/2)$, the more sensitive the bearing life will be to heating. This increase is the temperature jump between the raceways and the environment in the reference testing, where ε = 1/2. This value increases if there is poor cooling, higher rotational speed, greater radial force, or a higher friction coefficient. However, the effect of thermal dilation remains to be considered. Indeed, the greater heating of the raceways can reduce the radial clearance, which in turn increases the friction. The model presented can include this physical process if we replace ε with a new parameter that we will call ε^* in expressions (5.67) to (5.69). Because thermal dilation does not affect the deformation due to the load (although it will affect bearing life by increasing the load on the rolling elements), we define the new parameter as:

$$\varepsilon^* = \frac{1}{2} \frac{x_Q}{x^*} = \frac{x_Q}{2x_Q + x_0 - \beta(T(\varepsilon^*) - T_0)} \tag{5.70}$$

where β is a constant that depends on the thermal expansion coefficient and the dimensions. Using expressions (5.46), (5.47), (5.61) and (5.63), the above equation is rewritten as:

$$\varepsilon^*(\varepsilon)=\frac{\varepsilon}{1-\chi_\infty-\chi_f(1/2)\dfrac{I_1(\varepsilon^*(1/2),n)I_2(\varepsilon^*(\varepsilon),n)}{I_1(\varepsilon^*(\varepsilon),n)I_2(\varepsilon^*(1/2),n)}} \tag{5.71}$$

$$\chi_\infty=\frac{\beta(T_\infty-T_0)}{2x_Q+x_0} \tag{5.72}$$

$$\chi_f(1/2)=\frac{\beta\Delta T(1/2)}{2x_Q+x_0} \tag{5.73}$$

Equation (5.71) is an implicit equation that enables us to calculate $\varepsilon^* = \varepsilon^*(\varepsilon)$. Constants χ_∞ and $\chi_f(1/2)$ measure the fraction that the thermal expansions represent in the diametral clearance of the cold, loaded bearing. The first is caused by the heating of the bearing to machine temperature $(T_\infty - T_0)$, and the second by the heating due to friction $(\Delta T(1/2))$. Thus, bearing life can finally be written as:[17]

$$\frac{L_{10}(\varepsilon)}{L_{10}(1/2)}=\left(\frac{I_1(\varepsilon^*(\varepsilon),n)}{I_1(\varepsilon^*(1/2),n)}\left[1-2.52\cdot10^{-4}\Delta T(1/2)\left(\frac{I_1(\varepsilon^*(1/2),n)I_2(\varepsilon^*(\varepsilon),n)}{I_1(\varepsilon^*(\varepsilon),n)I_2(\varepsilon^*(1/2),n)}-1\right)\right]^{3.6}\right)^p \tag{5.74}$$

Because greater dilation means greater friction, and greater friction means greater thermal expansion, the phenomenon described by equation (5.71) might prove to be unstable. The presence of such instability is analytically corroborated because the denominator in equation (5.71) is canceled by a certain value of ε^*. Let ε_p be the value that cancels this denominator. In first approximation, the denominator in (5.71) can then be written as:

$$D(\varepsilon^*)=1-\chi_\infty-\chi_f(1/2)\frac{I_1(\varepsilon^*(1/2),n)I_2(\varepsilon^*,n)}{I_1(\varepsilon^*,n)I_2(\varepsilon^*(1/2),n)}$$

$$=\frac{dD(\varepsilon^*)}{d\varepsilon^*}(\varepsilon^*-\varepsilon_p)+0(\varepsilon^*-\varepsilon_p)^2 \tag{5.75}$$

$$D(\varepsilon_p)=0 \tag{5.76}$$

This approximation transforms equation (5.71) into a second-degree equation that can be solved:

$$\varepsilon^{*2} - \varepsilon^{*}\varepsilon_P - \frac{\varepsilon}{\dfrac{dD(\varepsilon_P)}{d\varepsilon_P}} \simeq 0 \Rightarrow \varepsilon^{*} \simeq \frac{\varepsilon_P}{2} \pm \sqrt{\left(\frac{\varepsilon_P}{2}\right)^2 + \frac{\varepsilon}{\dfrac{dD(\varepsilon_P)}{d\varepsilon_P}}} \quad (5.77)$$

Because the derivative $dD(\varepsilon_p)/d\varepsilon_p$ is negative, the discriminant becomes negative after a particular value of ε that we will call the critical value ε_C. The bearing therefore collapses due to its own heating if the initial radial clearance (obtained for the cold, loaded bearing) has a value of ε that exceeds the value ε_C given by:

$$\varepsilon_C = -\left(\frac{\varepsilon_P}{2}\right)^2 \frac{dD(\varepsilon_P)}{d\varepsilon_P} \quad (5.78)$$

To complete the model, it is necessary to estimate the value of exponents n and p. A geometric calculation shows that the depth of the deformed volume is proportional to the square of the width of the contact area. Thus, if the depth of the effective volume over which the load is supported is δ and its characteristic width is a, the deformed depth is $\delta \propto a^2$. The strain is proportional to $\delta / a \propto a$ and, consequently, the elastic behavior of the material establishes that the pressure exerted on the contact, which is proportional to the strain and the elastic modulus, is also proportional to a. The load exerted on the contact will be the pressure (proportional to a) multiplied by the contact area, which is a^2 for spherical contact and al for cylindrical contact, where l is the cylinder length. Consequently, the load supported by a contact is proportional to the volume of the affected material (the complete functional relationship is obtained from a Hertz stress analysis at a point contact between two elastic parts; for a case involving bearings, see Harris, 2001). It is therefore $Q \propto a^3$ for a ball bearing, and $Q \propto a^2$ for a roller bearing. Finally, given that $a \propto \delta^{1/2}$, we have $Q \propto \delta^{3/2}$ or $Q \propto \delta$, i.e. $n = 3/2$ for a ball bearing, and $n = 1$ for a cylinder bearing. This confirms the hypothesis used to write equation (5.49). The theoretical estimation of p is not immediate.[18] However, bearing operation and testing experience shows that $p = 3$ for a ball bearing, and $p = 4$ for a cylinder bearing.

Figure 5.6 shows the life of a bearing in three situations: a) for a cold bearing (the temperature of the raceways does not change), b) for a hot bearing without thermal expansions (temperature variations due to changes in friction only affect the loss of mechanical properties), and c) for a hot bearing with dilations (the average machine temperature and the friction produce dilations that reduce the radial clearance). As we can see, for the cold bearing, there is a radial clearance value that maximizes the lifetime. That value is found for the values $\varepsilon > 1/2$, meaning that for

Figure 5.6 **Dimensionless bearing lifetime as a function of diametral clearance**

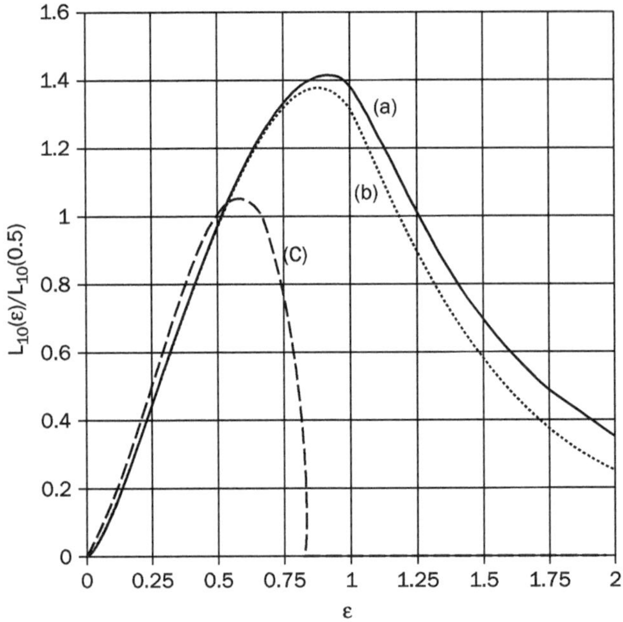

Note: Curve (a) takes into account the effect of diametral clearance on angular load distribution. Curve (b) includes the effect of friction on the loss of surface hardness. Curve (c) takes into account the effect of thermal expansions. The curves are obtained using the following constant values: a) $n = 3/2$, $p = 3$, $\Delta T(1/2) = 0$, $\chi_\infty = 0$ and $\chi_f(1/2) = 0$; b) $n = 3/2$, $p = 3$, $\Delta T(1/2) = 50°C$, $\chi_\infty = 0$ and $\chi_f(1/2) = 0$; and c) $n = 3/2$, $p = 3$, $\Delta T(1/2) = 50°C$, $\chi_\infty = 0.15$ and $\chi_f(1/2) = 0.15$

cold bearings, it is good for the radial clearance to be negative, i.e. with a certain degree of interference between the raceways and the rolling elements. However, for hot bearings, the interference present when cold can cause excessive interference when hot, to the point where the lifetime is reduced to zero for the clearance values that would produce a maximum lifetime when cold. For this reason, precision shaft guiding, which requires interference ($\varepsilon > 0.5$), requires an exquisite design and control of the lubrication, cooling and assembly system. Otherwise, there is a risk of significantly reducing the bearing life.

The reduction of the lifetime can be even greater, given that all variables intervening in the problem, n, p, $\Delta T(1/2)$, χ_∞, $\chi_f(1/2)$, and ε, are affected by noise and inaccuracies. In particular, ε depends on the value of the load (through x_Q) and diametral clearance x_0 obtained during assembly. In turn, x_0 depends on the clearance before assembly (depends on bearing production) and the interference fitting generated while mounting the bearing on the machine. Finally, the interference fittings during assembly depend on the tolerances (dimensional and geometrical) and surface finishes on the bearing housings. The random behavior of these variables means that the expected lifetime for a given reliability is also a random variable. The probability that such noise will reduce the bearing life to below the value preset during design can be calculated thanks to the FORM procedure explained in Section 5.4.

An approximate method such as FORM requires that the value of x_0 be determined from the production and assembly parameters. The radial clearance of a mounted bearing depends on the combination of tolerances between the shaft, housing and bearing. When a ring with an elastic modulus of E_2, a Poisson coefficient of ν_2, inner radius R_{2i} and outer radius R_{2o} is loaded with an interior pressure of P, the inner radius increases by the amount $\Delta R_{2i} = R_{2i}\dfrac{P}{E_2}\left[\dfrac{(R_{2o}/R_{2i})^2+1}{(R_{2o}/R_{2i})^2-1}+\nu_2\right]$. When a ring with an inner radius of R_{1i} and an outer radius of R_{1o} is loaded with an exterior pressure of P, the outer radius decreases by the amount $\Delta R_{1o} = R_{1o}\dfrac{p}{E_1}\left[\dfrac{(R_{1o}/R_{1i})^2+1}{(R_{1o}/R_{1i})^2-1}-\nu_1\right]$. When two rings are mounted concentrically with an interference of $I = 2(R_{1o} - R_{2i})$, the size of the outer ring will increase and that of the inner ring will decrease in order to reach a situation in which both rings are subjected to the same pressure, P, and have the same radius, $R_{2i} + \Delta R_{2i} = R_{1o} - \Delta R_{1o}$. Consequently, the interference fitting is $I = 2(\Delta R_{2i} + \Delta R_{1o})$. This expression enables us to calculate the pressure, P, to which the contact surface is subjected. By performing this analysis for the housing and the outer ring of the bearing, and for the shaft and the inner ring of the bearing, we can establish the new diametral clearance around the bearing once it is installed. This diametral clearance will depend on the radius changes associated with the raceways. In the particular case in which the housing, the bearing and the shaft are made of the same material, the shaft is solid, and the housing has much larger exterior dimensions than the bearing, the decrease in the diametral clearance is determined by $\Delta = I\left(\dfrac{R_S}{R_1}+\dfrac{R_2}{R_H}\right)$. Here, R_S is the shaft diameter, R_H is the diameter of the bearing housing in the case, R_1 is the outer diameter of the inner ring (inner raceway), and R_2 is the inner diameter of the outer ring (outer raceway).[19]

In addition to seriously affecting the service life of the bearing, the radial clearance is also related to other functional requirements and constraints on the problem, such as shaft guiding accuracy, ease of assembly and disassembly, allowing axial displacement of the shaft,[20] rigidity of the contact (especially for ball bearings, as their rigidity depends strongly on the load, $n = 1.5$)[21], compensation for wear and tear and settling with use, etc.

This section shows that the parameter ε relates the reliability and guiding precision of the shaft (functional requirements) to the following design parameters: load, rotation regime, lubricant, temperatures, materials and assembly interferences (which, in turn, depend on the surface finishes and shape tolerances). It is an interesting exercise, which will be left for the reader to solve, to use Axiomatic Design to create a bearing that eliminates most of these dependencies. It is also interesting for the reader to use the robust design equation from Metric Design to obtain the radial clearance value that provides optimum guiding precision and shaft reliability.

5.7 Notes

1. The Appendix explains some of the concepts required to specify reliability as a function of component life. For an introduction to reliability, please see Tobias and Trindade (1986), Cruse (1997) and Thompson (1999).
2. For example, on a precision positioning system, the maximum deflection on a cantilever beam subjected to end load P must not exceed a particular value. If L is the length of the beam, I is the moment of inertia for the section, E is the elastic modulus, and δ is the maximum deflection admitted by the customer, the failure criterion will be $\delta < PL^3/(3EI)$. Note that this failure criterion also leads to the functional requirement that describes the customer's need. Indeed, the acceptance interval is $[-\delta, \delta]$ and the response is $PL^3/(3EI)$. For this reason, expression (3.13) and the corresponding transfer function enable us to write the functional requirement $y = PL^3/(3EI\delta)$, with the acceptance interval $[-1,1]$, as a function of the operation and design parameters P, L, E, I and δ. According to (3.28), the probability of success, i.e. the reliability, will be the probability $\Pr[y \in [-1,1]]$. However, although the transfer function takes into account the variation produced by the noise present in the operation and design parameters, it does not consider other possible failure modes such as the impact of foreign elements on the beam, corrosion, flow, vibrations in the support, etc. Each of these events is a failure mode with a probability of occurrence. The event with the highest probability of occurrence will be the critical failure mode, and will establish the order of magnitude of the reliability.
3. Note that, in general, function g_i is generic because it may or may not coincide with the definition of the functional requirement.

4. If the expected probabilities of failure are much lower than one, it is possible to approach the reliability using the expression $R = 1 - \sum_{i=1}^{FMA} F_i$. The error involved in using this expression is less than $0.5\max(F_iF_j)FMA^2$.
5. This expression for the error can be found in Cruse (1997). This reference describes how to reduce the error through the appropriate selection of a probability distribution function for the point sampling. These techniques for calculating the probability integral are known as the Monte Carlo method.
6. A reliability of 10^{-3}, typical for components in the automobile industry and other industrial sectors, means accepting a failure in one of each 1000 components. In the aerospace industry, the reliabilities sought for components are even higher, in the order of 10^{-6} and 10^{-9}. For example, in 2008 the reliability of fixed-wing aircraft with a maximum takeoff weight of 2250 kg was roughly five fatal accidents for each 10 million takeoffs. If we assume that an aircraft has about 100 critical failure modes, the probability of each failure mode must be in the order of $5 \cdot 10^{-9}$.
7. The safety coefficient of an elevator is in the order of 14, while that of a missile is in the order of 1.1.
8. An increase in safety factors usually increases system mass, volume and cost. On an aircraft, increasing mass reduces the payload and increases operating costs. In an electronic circuit, the safety factors increase the volume and cost.
9. Axiomatic Design and Metric Design use the correct selection of the design matrix to increase reliability. In the case of the archer discussed in Chapter 4, increasing the exit velocity of the projectile reduces the influence of the distribution tails associated with all of the mechanisms (and operations) for tensing the bow (selection, positioning, and tensing the rubber bands). If we consider the previous chapters, the design strategy for obtaining the maximum reliability involves reducing the sensitivity of the response to noise. The minimum sensitivity is usually obtained when the range of variation permitted for certain parameters is displaced towards excessively high values. In this sense, this strategy is equivalent to increasing the safety factors.
10. The FORM method presented has several limitations. 1) The variables that constitute vector x must be independent and normal. If they are not, a variable change must be established to ensure it. 2) The approximation used is first-order. If the curvature is important in a small neighborhood of the most probable point, the algorithm cannot converge. Hypersurfaces with a large curvature or very high reliabilities can produce errors in the order of the failure probability itself. In addition, even though the components of vector x are statistically independent and normal, distribution g might not be normal. 3) If there are several points that give the minimum distance to the origin locally, the one reached will depend on the initial point chosen. Cruse (1997) proposes different solutions to these limitations. Nonetheless, aside from the purely mathematical exceptions, the current capacities of the processes used in today's engineering practices cause probability distributions to decline quickly enough when moving away from the origin (always in reduced variables) so that only the hypersurface shape near the origin

matters. Because this shape is set by the hyperplane tangent to the hypersurface, in first approximation, FORM is valid in 90 per cent of the cases presented (Rackwitz, 2001). Grandhi and Wang (1999) published a review of the approximate methods. Du and Sudjianto (2004) and Ba-abbad et al. (2006) introduced variations of the method proposed. Lian and Kim (2006) and Cruse and Sankaran (1997) presented designs of aeronautical components based on reliability criteria.

11. Because this is a first-order approximate method, the number of iterations required to reach the desired precision depends on the curvature around the maximum probability point. It is an interesting exercise to check whether the failure criterion $g = AY/Q - 1 < 0$ provides the same probability of failure and the same sensitivities as the one used. To avoid complications due to the curvature of $g(x) = 0$ in the iterative process, it is advisable to rewrite the failure criterion in the most linear way possible.
12. If the area of the minimum cross-section of the valve is A, the part density is ρ, its elastic modulus is E, and the impact velocity is u, then an estimation of the force generated in that section is $Q = 2\sqrt{\rho E A}u$. Moreover, the counter-pressure, P, the area where it acts, A_V, the initial lift of the valve, h, and its mass, m, determine the impact velocity using the expression $u = \sqrt{2PA_V h / m}$. The failure criterion would be $g = Y - Q/A = Y - 2\sqrt{2EPA_V h\rho / m}$. Note that with this new failure criterion, increasing the valve area is counterproductive.
13. In equation (5.41), L is used as a time variable because lifetime is usually measured in millions of load cycles. Nonetheless, for systems where the load cycle is not clearly defined, expression (5.41) is valid if we simply replace L with the time, t. For quick sizing, the reliability is usually set to 0.9. Equation (5.41) was prepared so that a device subjected to a constant load $dQ(L)/dL = 0$ with reliability $R = 0.9$ has a lifetime that behaves according to potential law $L = (C/Q)^p$, which is the formula used to estimate the lifetime of many electronic and mechanical devices, for example, lasers (Fukuda, 1995), LEDs (Levada et al., 2005), and bearings (Harris, 2001). The exponent p is usually called the Coffin–Manson exponent, and must be obtained experimentally for virtually all devices (which is what makes the proposed method semi-empirical). Nevertheless, this exponent is usually the same if the material and failure mode are also the same. For example, Blish (1997) provides a table with the typical exponents for different fragile fracture modes. The validity of the model given in (5.41) and the calculation of some exponents for thermal shock and thermal fatigue are discussed by Benavides (2010).
14. One case where the distinction is quite clear is that of a metal bearing (manufactured with steel hardened to 60 to 64 Rockwell C) with a surface fatigue failure mode (Harris, 2001). In this case, $s = 1.12$ for all designs and sizes, $p = 3$ for all designs where the rolling elements are balls, regardless of size ($p = 4$ for all designs where the rolling elements are cylinders, regardless of size), and C varies greatly with the bearing design and size.
15. There are more failure modes that lead to the loss of position of the contact surfaces. For example, the breakage of one of the rings into two parts is another failure mode. It is an interesting exercise to list all possible failure

modes and classify them by: 1) the possibility of eliminating them by incorporating appropriate systems, and 2) how critical they are.

16. Expressions (5.56) and (5.57) are useful approximations for a sensitivity study, but the data and information supplied by the bearing manufacturer must be used for a detailed design.
17. Comparison point $\varepsilon = 1/2$ produces a radial clearance equal to $\varepsilon^*(1/2) = 1/2/(1 - \chi_\infty - \chi_f(1/2))$ in operation (while hot).
18. This problem is discussed by Harris (2001).
19. Due to the crests and valleys that are always present on such surfaces, the surface treatment modifies the value of the expected interference. The decrease in the (radial) interference between two surfaces can change 2 to 5 μm for accurately polished surfaces, 6 to 14 μm for gently machined surfaces, 10 to 24 μm for reamed surfaces, and 24 to 48 μm for normally machined surfaces. However, the actual value of the decrease in interference depends on the rotational condition of the load, magnitude of the load, internal bearing clearance, temperature, materials, shaft and housing designs, etc.
20. Axiomatic Design, Section 3.7, showed that the best way to attach a shaft to a machine is with two bearings. Both fix the shaft radially, but only one fixes it axially. This is particularly important if the shaft is subject to strong thermal expansions. There are two possibilities for allowing free movement of the shaft: 1) the axial displacement can be absorbed by the bearing itself (cylindrical or needle roller bearings), and 2) it can be absorbed by an axial displacement between the raceway and the shaft or housing in the oil pan (ball bearings mounted without interfering with the shaft or housing). If the latter solution is adopted, the raceway that must be kept free axially is the one subject to a fixed load (the direction of the load does not change with respect to the raceway).
21. Errors by inertial navigation gyroscopes depend on the rigidity of the shaft supports. The selection of the radial rigidity, through an interference preload of the rolling elements, is essential for ensuring correct operation.

6

Entropy-based design

Abstract: Corollary 2 from Axiomatic Design and the reduction of Quality Loss specify that the number of functional requirements must be minimal: this minimum number is one. The designer must therefore seek the path that resolves only one functional requirement on each level of the design hierarchy. If the designer manages to maintain the ideal situation (in which each design step resolves a single functional requirement), the Independence Axiom is met automatically, and therefore adds no information. This chapter proves two theorems that help channel decision-making in this case. The two theorems proposed are the Broad Tolerance Theorem and the Linearity Theorem. Both are applied to the design of a fuel supply system, showing how they completely freeze a conceptual design. Finally, the chapter concludes by relating design activities to entropy generation and reduction.

Key words: tolerance, linearity, conservation laws, entropy generation, information generation, design process.

6.1 The Minimum Tolerance Theorem

Broad Tolerance Theorem: Of all the designs that verify the list of constraints and the Independence and Information axioms, those with broad tolerances are better than those with strict tolerances.

Proof: Corollary 6 from Axiomatic Design and the reduction of Quality Loss require that the acceptance intervals be made as broad as possible. Given that the design parameters on one level of the design hierarchy become the functional requirements on the next level, establishing strict tolerances for the design parameters contradicts Corollary 6.

Theorem of the Minimum Number of Tolerances: Of all the designs that verify the list of constraints and the Independence and Information Axioms, those with the smallest number of tolerances are the best.

Proof: This is an immediate consequence of the Broad Tolerance Theorem. When the tolerance for a design parameter is high enough, so that the tolerance is much higher than the natural variability (produced by the noise affecting that design parameter), the tolerance can be eliminated without causing any changes to the design.

6.2 The Linearity Theorem

Let l be a variable belonging to the Space of Definition for the Needs, and m a variable belonging to the Space of Definition for the Solution. Following the nomenclature introduced in Chapter 2, the acceptance interval for variable l is $[\underline{l},\overline{l}]$, and the variation interval for variable m is $[\underline{m}, \overline{m}]$. Both variables can be encoded according to expressions

$$l = \frac{\overline{l} + \underline{l}}{2} + y\frac{\overline{l} - \underline{l}}{2} \tag{6.1}$$

$$m = \frac{\overline{m} + \underline{m}}{2} + x\frac{\overline{m} - \underline{m}}{2} \tag{6.2}$$

These expressions define the dimensionless variables x and y (which must verify $x \in [-1, +1]$ and $y \in [-1, +1]$). According to Axiomatic Design, the best design is the one with a single functional requirement (see Axiomatic Design Corollary 2) and a single design parameter (Axiomatic Design Theorem 4). Therefore, the synthesis operator responsible for generating a solution must propose a solution with the transfer function

$$l = f(m) \tag{6.3}$$

This transfer function automatically meets the Independence Axiom. By replacing (6.1) and (6.2) in (6.3), we obtain

$$y = \frac{2}{\overline{l} - \underline{l}}\left[f\left(\frac{\overline{m} + \underline{m}}{2} + x\frac{\overline{m} - \underline{m}}{2}\right) - \frac{\overline{l} + \underline{l}}{2}\right] \tag{6.4}$$

This new transfer function enables us to identify x as a new variable from the Space of Definition for the Solution, and y as a variable belonging to the Space of Definition for the Response. Obviously, the design will be accepted when the response margin $[\underline{y}, \overline{y}] = [\inf[y([-1,+1])],$

sup[y([−1,+1])]] for variable y verifies $[\underline{y},\overline{y}] \subseteq [-1,+1]$, a condition that ensures compliance with the Information Axiom.

Transfer function (6.4) should be rewritten as:

$$y=\frac{2}{\overline{l}-\underline{l}}f\left(\frac{\overline{m}+\underline{m}}{2}\left(1+x\frac{\overline{m}-\underline{m}}{\overline{m}+\underline{m}}\right)\right)-\frac{\overline{l}+\underline{l}}{\overline{l}-\underline{l}} \tag{6.5}$$

Given that $x \in [-1,+1]$, when the condition $\overline{m}-\underline{m} \ll \overline{m}+\underline{m}$ is verified, the above expression can be developed as a Taylor series:

$$y=\frac{2}{\overline{l}-\underline{l}}\left[f\left(\frac{\overline{m}+\underline{m}}{2}\right)+f'\left(\frac{\overline{m}+\underline{m}}{2}\right)x\frac{\overline{m}-\underline{m}}{2}+\frac{1}{2}f''\left(\frac{\overline{m}+\underline{m}}{2}\right)\right. \tag{6.6}$$

$$\left.\times\left(x\frac{\overline{m}-\underline{m}}{2}\right)^2+0\left(x\frac{\overline{m}-\underline{m}}{\overline{m}+\underline{m}}\right)^3\right]-\frac{\overline{l}+\underline{l}}{\overline{l}-\underline{l}}$$

Because there is only one functional requirement and a single design parameter, the design matrix contains a single element, which we will call A. This element is the linear term in expression (6.6), whose definition coincides with the one adopted in matrix (3.8). Transfer function (6.6) can therefore be written as

$$y=\alpha+Ax+\beta x^2 \tag{6.7}$$

$$\alpha=\frac{2}{\overline{l}-\underline{l}}f\left(\frac{\overline{m}+\underline{m}}{2}\right)-\frac{\overline{l}+\underline{l}}{\overline{l}-\underline{l}} \tag{6.8}$$

$$A=f'\left(\frac{\overline{m}+\underline{m}}{2}\right)\frac{\overline{m}-\underline{m}}{\overline{l}-\underline{l}} \tag{6.9}$$

$$\beta=f''\left(\frac{\overline{m}+\underline{m}}{2}\right)\frac{\overline{m}-\underline{m}}{\overline{l}-l}\frac{\overline{m}-\underline{m}}{4} \tag{6.10}$$

If we assume that x is a random variable uniformly distributed in interval [−1,+1], the device's response will have the following mean and variance

$$E[y]=\alpha+\frac{\beta}{3} \tag{6.11}$$

$$E\left[y^2-E[y]^2\right]=\frac{A^2}{3}+\frac{4\beta^2}{45} \tag{6.12}$$

The non-linear term β in expression (6.7) appears in both expressions, and therefore produces two effects: 1) it changes the center of the response interval, and 2) it changes the width of the response interval. If β is large enough, the condition $[\underline{y},\overline{y}] \subseteq [-1,+1]$ will no longer be valid, and the

Information Axiom will not be met. One way to cause $[\underline{y},\overline{y}] \subseteq [-1,+1]$ to be verified is by imposing the conditions

$$E[y] = \alpha + \frac{\beta}{3} = 0 \tag{6.13}$$

$$E\left[y^2 - E[y]^2\right] = \frac{A^2}{3} + \frac{4\beta^2}{45} \ll \frac{1}{3} \tag{6.14}$$

These two equations impose two new needs on the design[1], whose transfer functions are obtained by replacing (6.8), (6.9) and (6.10) in (6.11) and (6.12):

$$y_1 = \frac{2}{\overline{l} - \underline{l}} f\left(\frac{\overline{m} + \underline{m}}{2}\right) - \frac{\overline{l} + \underline{l}}{\overline{l} - \underline{l}} + f''\left(\frac{\overline{m} + \underline{m}}{2}\right) \frac{\overline{m} - \underline{m}}{\overline{l} - \underline{l}} \frac{\overline{m} - \underline{m}}{12} \tag{6.15}$$

$$y_2 = \frac{1}{3} f'\left(\frac{\overline{m} + \underline{m}}{2}\right)^2 \left(\frac{\overline{m} - \underline{m}}{\overline{l} - \underline{l}}\right)^2 + \frac{4}{5} f''\left(\frac{\overline{m} + \underline{m}}{2}\right)^2 \times \left(\frac{\overline{m} - \underline{m}}{\overline{l} - \underline{l}}\right)^2 \left(\frac{\overline{m} - \underline{m}}{12}\right)^2 \tag{6.16}$$

In these new design equations, the design parameters are now $\underline{m}$ and $\overline{m}$. We can see that the associated design matrix is size 2 and coupled:

$$\begin{bmatrix} \dfrac{\partial y_1}{\partial \underline{m}} & \dfrac{\partial y_1}{\partial \overline{m}} \\ \dfrac{\partial y_2}{\partial \underline{m}} & \dfrac{\partial y_2}{\partial \overline{m}} \end{bmatrix} \neq \begin{bmatrix} X & 0 \\ 0 & X \end{bmatrix} \tag{6.17}$$

One way of improving the design is by uncoupling the matrix in (6.17). As we can see in equations (6.15) and (6.16), the condition $\underline{m} = \overline{m}$ allows immediate compliance with (6.13) and (6.14). However, this result contradicts the Broad Tolerance Theorem. Therefore, in order to impose compliance with Axiom 1 when the tolerances cannot be null, we will take two new design parameters, m and $\delta \neq 0$, according to:

$$\underline{m} = m - \delta \tag{6.18}$$

$$\overline{m} = m + \delta \tag{6.19}$$

With these new parameters, equations (6.15) and (6.16) would be:

$$y_1 = \frac{2}{\overline{l} - \underline{l}} f(m) - \frac{\overline{l} + \underline{l}}{\overline{l} - \underline{l}} + \frac{1}{6} f''(m) \frac{2\delta^2}{\overline{l} - \underline{l}} \tag{6.20}$$

$$y_2 = \frac{1}{3}\left[f'(m)^2 + \frac{\delta^2}{15} f''(m)^2 \right]\left(\frac{2\delta}{\overline{l} - \underline{l}} \right)^2 \tag{6.21}$$

The design matrix can be obtained as follows:

$$m = m_0(1 + \varepsilon_1), \text{ where } \varepsilon_1 \ll 1 \tag{6.22}$$

$$\delta = \delta_0(1 + \varepsilon_2), \text{ where } \varepsilon_2 \ll 1 \tag{6.23}$$

Using ε_1 and ε_2 as new design parameters, the resulting design matrix is

$$\begin{bmatrix} \dfrac{2f'(m_0)m_0}{\overline{l} - \underline{l}} + \dfrac{1}{3}\dfrac{f'''(m_0)m_0\delta_0^2}{\overline{l} - \underline{l}} & \dfrac{2}{3}\dfrac{f''(m_0)\delta_0^2}{\overline{l} - \underline{l}} \\ \dfrac{2}{3}\left(\dfrac{2\delta_0}{\overline{l} - \underline{l}} \right)^2 \left[f'(m_0) + \dfrac{\delta_0^2}{15} f'''(m_0) \right] f''(m_0)m_0 & \dfrac{2}{3}\left(\dfrac{2\delta_0}{\overline{l} - \underline{l}} \right)^2 \left[f'(m_0)^2 + \dfrac{2\delta_0^2}{15} f''(m_0)^2 \right] \end{bmatrix} \tag{6.24}$$

As indicated in (6.17), this design matrix is coupled. Nevertheless, the terms outside of the main diagonal are canceled when the second derivative of transfer function (6.3) is null. If we had retained more terms in the development of Taylor series (6.6), the elements in the design matrix would depend on higher-order derivatives. To eliminate them, it would also be necessary to cancel the third, fourth and higher derivatives. Canceling all derivatives higher than first-order is equivalent to stating that the transfer function is linear. Therefore, compliance with the Independence Axiom in (6.24) enables us to state the following theorem:

Linearity Theorem:[2] Linear designs are better than non-linear designs.

An alternative way to state this theorem is:

Linearity Theorem (Statement 2): Linear design parameters are better than non-linear parameters. In other words, when there is more than one design parameter affecting a particular functional requirement, the one that produces the most linear variation of the functional requirement should be chosen.

This theorem explains why, in the faucet example described in Chapter 3, parameter *D* selected the outlet areas as adjustment parameters instead of the pressures. It also explains why Robust Design selected the position as an adjustment parameter instead of the angle in the archer example described in Chapter 4.

6.3 Example application: conceptual design of a fuel supply system for gasoline engines

This example shows the power of using the Minimum Tolerance and Linearity Theorems jointly.[3] It also illustrates how the Axioms and Theorems deduced throughout the book guide the conceptual design process for the system until the final design is frozen.

6.3.1 Stating the problem

Motivation: To provide the fuel required by a forced-ignition piston engine (gasoline engine).

Space of definition for the needs: The customer specifies that the fuel required by the engine is

$$\dot{m}_f = (1 \pm \varepsilon_f)\dot{m}_a(\phi,n,T_a,T_e,P_a,\alpha_1,\ldots,\alpha_m)F_r(\phi,n,T_a,T_e,P_a,\beta_1,\ldots,\beta_n)F_{st} \quad (6.25)$$

Thus, the customer has established that the fuel required is the product of the discharge air and the richness of the mixture. Both the discharge air and the richness are functions that depend on the point of operation and the detailed design of the engine. For this reason, the fuel required by the engine depends on throttle position ϕ, rotation regime n, ambient temperature T_a, engine temperature T_e, intake pressure P_a, stoichiometric richness F_{st}, design and operation parameters $\alpha_1, \ldots, \alpha_m, \beta_1, \ldots, \beta_n$ defining the geometry of the intake manifold, the valves, and the type of operation desired (maximum power, minimum consumption, or minimum pollution), among others. The customer also specifies a relative error of ε_f limiting the maximum variations admissible for the fuel discharge.

In the space of definition for the needs, we can define the acceptance interval as $\left[\underline{\dot{m}_f}, \overline{\dot{m}_f}\right]$, where the lower and upper limits of the interval are functions (6.26) and (6.27) obtained from (6.25):

$$\underline{\dot{m}_f} = (1-\varepsilon_f)\dot{m}_a(\phi,n,T_a,T_e,P_a,\alpha_1,\ldots,\alpha_m)F_r(\phi,n,T_a,T_e,P_a,\beta_1,\ldots,\beta_n)F_{st} \quad (6.26)$$

$$\overline{\dot{m}_f} = (1+\varepsilon_f)\dot{m}_a(\phi,n,T_a,T_e,P_a,\alpha_1,\ldots,\alpha_m)F_r(\phi,n,T_a,T_e,P_a,\beta_1,\ldots,\beta_n)F_{st} \quad (6.27)$$

As we can see, the acceptance limit depends on the engine's point of operation and its detailed design.

6.3.2 The design process

6.3.2.1 Selecting the functional requirements

Following the Axiomatic Design methodology, the first step in the design process involves reformulating the list of needs as a list of functional requirements (Corollary 2 recommends a list with a single item, if possible) and a list of constraints. In this problem, the set of needs is comprised of the acceptance intervals obtained at each point of operation and design. Obviously, they are all coupled through the point of operation and design, and form an infinite set (there is one for each value of the engine's points of operation and design). These intervals therefore do not constitute a minimum and independent set of needs, i.e. they do not constitute a set of functional requirements. To extract the set of functional requirements, we select one of those intervals. The choice of interval is arbitrary; consequently, the choice of upper and lower acceptance limits defining the interval is also arbitrary. However, when we set the engine's point of operation and design, the interval will no longer be arbitrary. For this reason, a constraint arises: the system must be capable of changing the response. In other words, except for a translation and a scale factor, all acceptance intervals must be accessible to the system. Thus, instead of using functions (6.25) to (6.27), set by the customer, to define the design problem, we will use the following:

List of needs: 1) Provide any fuel flow value, 2) convert that value into the one required by the engine's point of design and operation with the precision specified at that point.

From the above, we deduce the following list of functional requirements and constraints.

List of functional requirements: An arbitrary fuel flow rate.

List of constraints: The device must be capable of modifying the flow and ensuring the precision required at each point of design and operation.

We will condense the formulation of the problem using the nomenclature $\forall_T x$, which means any value of x affected by a tolerance. On the other hand, the sign $\forall x$ means any value of x and any value of its precision. Thus, $\forall_T \dot{m}_f$ means any fuel flow value, where the precision cannot be arbitrary.[4]

The main value added by the new formulation of the problem is the versatility of the solution we will obtain. The conversion of the customer's need given in (6.25) (specific to a particular customer and engine) has become the functional requirement $\forall_T \dot{m}_f$ (independent of the customer and independent of the engine).[5]

6.3.2.2 Selecting the solution

The functional requirement $\forall_T \dot{m}_f$ has been selected in the complete absence of a solution. The main advantage of not adopting any solution when formulating the problem is the elimination of any possible design parameter inherited from existing solutions (which may not be the best).[6] In the complete absence of a solution, the formulation of the motivation is as general as: 'take the fuel that is outside of the engine and introduce an amount per unit of time equal to $\forall_T \dot{m}_f$'. The first level of the design hierarchy must therefore determine what boundary in the engine the fuel must cross.

First level

On this level of the design hierarchy, there is only one functional requirement: to pass through the surface of the engine in order to reach the air that circulates through it. Thus, this approach meets Axiomatic Design Corollary 2.

In general terms, there are two boundaries that the fuel can cross: the intake manifold and the working chamber (made up of the piston, cylinder and cylinder head).[7] The configuration of each of these surfaces depends on the design parameters specifying the geometry of the engine, and the needs that the surfaces must satisfy. Thus, the surface that affects the highest number of functional requirements will be the worst solution. The cylinder head, piston and cylinder ensure structural integrity, cooling, intake, exhaust, combustion, power output, etc. However, the manifold only ensures intake. According to Axiomatic Design, the intake manifold is therefore a better option than the chamber.

Decision adopted: The fuel will pass through the intake manifold.

Second level

On this level of the design hierarchy, it is known that the fuel will pass through the surface of the intake manifold. It is therefore known that the discharge pressure for the system we design is the intake manifold pressure, P_a. This pressure is a function that depends on the engine's point of operation and design. The laws of conservation of energy and mass specify that the fuel flow passing through the surface of the intake manifold is

$$\dot{m}_f = \sqrt{2\rho_f (P_f - P_a)} A_f C_D(\mathrm{Re}, \gamma_1, \ldots, \gamma_p) \tag{6.28}$$

This physical law establishes that the fuel flow depends on fuel density ρ_f, fuel supply pressure P_f (to be determined), intake manifold pressure (determined by the engine configuration), characteristic fuel passage area A_f (to be determined), and function C_D, which covers everything not adequately modeled by the above terms (also to be determined). Because the details of how the fuel will pass through the surface are not yet known, the function $C_D(\mathrm{Re},\gamma_1,\ldots,\gamma_p)$, which specifies the discharge coefficient of the device, is completely unknown. This function depends on the Reynolds number and all of the design parameters $\gamma_1,\ldots,\gamma_p$, which will be known once the solution becomes more detailed. Due to the arbitrarity of function C_D, this equation is the same for all designers who have reached this point.

Equation (6.28) establishes that for any flow value, the designer must specify any value of the parameters on which it depends. In other words, (6.28) can be rewritten as

$$\forall_T \dot{m}_f = \sqrt{2\rho_f(\forall_T P_f - P_a)}\,\forall_T A_f\,\forall_T C_D(\mathrm{Re},\gamma_1,\ldots,\gamma_p) \qquad (6.29)$$

Equation (6.29) shows that to find any flow value with a particular tolerance, the designer can impose any area value (with a tolerance inherited from the customer's), any pressure value, or any variation of the system details. On this level of the design hierarchy, the issue the designer must resolve is which parameter is the best to vary. In other words, he must choose between $\forall_T P_f$, $\forall_T A_f$ and $\forall_T C_D$.

The design philosophy established in this book leads to the following information: 1) the Linearity Theorem selects the area rather than the pressure, and 2) the Minimum Tolerance Theorem selects the area rather than the discharge coefficient. (Knowing the discharge coefficient with sufficient precision implies knowing its arguments, also with sufficient precision). While the discharge coefficient introduces many tolerances, the area could introduce only one: the one affecting its own value.[8] Consequently, the decision adopted is:

Decision adopted: The area will be used, with a single tolerance, to vary the flow.

Third level

On this level of the design hierarchy, the designer must formulate a solution capable of ensuring the new functional requirement established on the previous level. In other words, he must formulate a solution capable of achieving $\forall_T A_f$. In everyday language, the design problem on this level would be to find a system capable of generating any passage area value with a given precision. Formulating solutions is an activity that falls outside the scope of the methodology presented in this book.

However, this methodology provides sufficient clues to lead the synthesis operator to create the best solution directly.[9]

The Theorem of the Minimum Number of Tolerances specifies that our solution can only make use of one tolerance. It occurs to the designer that with a single tolerance, the only thing he can produce is a hole. To vary the area of the hole, he must cover it, but nothing he uses to cover it can have tolerances. (The single tolerance permitted was already devoted to specifying the diameter of the bore hole.) For this reason, he designs a plug consisting of a needle with symmetry of revolution (this meets Corollary 5), but with an unspecified shape. (The tip may be a cone with any angle, a paraboloid, a sphere, or any other shape defined by the production experts.) The only requirement is that the closure of the needle over the hole be fueltight.

Because the details of the needle tip are not specified, it is also not possible to specify the outlet area for intermediate positions in the area between zero and the maximum value. To find all of the intermediate values (as a function of the lifting of the needle), we would need more than one tolerance. Having reached this point, the argumentation imposed by both theorems has led us to a variable area which must be either zero or the maximum value (a discontinuous, non-stationary process). However, equation (6.28) implicitly assumed that the solution must be stationary. To resolve this contradiction, the designer must replace equation (6.28) with the following:

$$\dot{m}_f = \frac{n}{60}\int_0^{\Delta t}\sqrt{2\rho_f(P_f - P_a(t))}A_f(t)C_D(\mathrm{Re},\gamma_1,\ldots,\gamma_p,t)\,dt \tag{6.30}$$

This equation assumes that the required fuel flow is obtained by providing discrete amounts of fuel. (To simplify, we have assumed that it is provided once with every turn of the crankshaft.) Let t_1 be the characteristic time for the duration of the opening transient, and t_2 the characteristic time for the duration of the closing transient. Outside of the transients, the area is either zero or the maximum value, and the discharge coefficient is constant. To consider the opening and closing transients, equation (6.30) can be written as

$$\begin{aligned}\dot{m}_f = {} & \frac{n}{60}\int_0^{t_1}\sqrt{2\rho_f(P_f - P_a(t))}A_f(t)C_D(\mathrm{Re},\gamma_1,\ldots,\gamma_p,t)\,dt \\ & + \frac{n}{60}A_f C_D(\mathrm{Re},\gamma_1,\ldots,\gamma_p)\int_{t_1}^{\Delta t}\sqrt{2\rho_f(P_f - P_a(t))}\,dt \\ & + \frac{n}{60}\int_{\Delta t}^{\Delta t+t_2}\sqrt{2\rho_f(P_f - P_a(t))}A_f(t)C_D(\mathrm{Re},\gamma_1,\ldots,\gamma_p,t)\,dt\end{aligned} \tag{6.31}$$

Compliance with the Theorem of the Minimum Number of Tolerances in (6.31) requires that the opening and closing transients, during which the hole is covered or uncovered, be negligible. (Otherwise, we would have to know the intermediate values for the outlet area as a function of the time with sufficient precision.) The design meets the Theorem of the Minimum Number of Tolerances if the characteristic time of these transients is much less than Δt, i.e. if

$$t_1 \ll \Delta t \tag{6.32}$$

$$t_2 \ll \Delta t \tag{6.33}$$

These two design equations[10] indicate that any value of t_1 and t_2 that is much lower than Δt is valid during design. For this reason, t_1 and t_2 are values that do not require tolerances, or for which the pertinent tolerances are so broad that they can be obviated. In this situation, the contribution of the transients to the integral in (6.31) can be disregarded, resulting in

$$\dot{m}_f = \frac{n}{60} A_f C_D(\mathrm{Re}, \gamma_1, \ldots, \gamma_p) \int_0^{\Delta t} \sqrt{2\rho_f (P_f - P_a(t))}\, dt \tag{6.34}$$

The second level of the design hierarchy imposed the condition $\forall_T \dot{m}_f$ on (6.28). However, this level of the design hierarchy has modified (6.28), replacing it with (6.34). By imposing $\forall_T \dot{m}_f$ on (6.34), we obtain

$$\forall_T \dot{m}_f = \frac{n}{60} A_f\, C_D(\mathrm{Re}, \gamma_1, \ldots, \gamma_p) \int_0^{\forall_T \Delta t} \sqrt{2\rho_f (\forall_T P_f - P_a(t))}\, dt \tag{6.35}$$

Because the area only takes two values on this level, neither the area nor the discharge coefficient can be chosen to ensure $\forall_T \dot{m}_f$.[11] The same argumentation used on the second level of the design hierarchy would eliminate the possibility of choosing $\forall_T P_f$. Thus, only $\forall_T \Delta t$ can be chosen as a solution. However, this variable is the upper limit of an integral. In general, the relationship between $\dot{m}_f$ and Δt is not linear. Having reached this point, the Linearity Theorem would not be met, and we would have a poor design. Again, the way to force compliance with the Linearity Theorem is by imposing the design equation

$$P_f - P_a(t) = \Delta P \tag{6.36}$$

This equation requires that the difference between the fuel supply pressure and the pressure in the intake manifold be constant and equal to ΔP. This value must be fixed by the appropriate tolerance. Thus, (6.36) turns (6.35) into

$$\dot{m}_f = \frac{n}{60} \sqrt{2\rho_f \Delta P}\, A_f\, C_D(\mathrm{Re}, \gamma_1, \ldots, \gamma_p) \Delta t \tag{6.37}$$

Decisions adopted:[12] The outlet area consists of a bore hole (a hole with a diameter specified by a particular tolerance) closed by a needle manufactured with symmetry of revolution. The control parameter ensuring the condition $\forall_T \dot{m}_f$ is Δt. In addition, design conditions (6.32), (6.33) and (6.36) must be met.

Fourth level

This level must ensure compliance with conditions (6.32), (6.33), and (6.36), established on the previous level.

Compliance with (6.32) and (6.33)

Compliance with design equations (6.32) and (6.33) requires the installation of devices capable of opening and closing the outlet area in times much shorter than the times it will remain open. Because the area is open for a time that must verify $\forall_T \Delta t$, equations (6.32) and (6.33) are written as

$$t_1 \ll \forall_T \Delta t \tag{6.38}$$

$$t_2 \ll \forall_T \Delta t \tag{6.39}$$

Given the generality of $\forall_T \Delta t$, it is possible to find values of Δt as small as we want. Therefore, the only possible solution to (6.38) and (6.39) is

$$t_1 \to 0 \tag{6.40}$$

$$t_2 \to 0 \tag{6.41}$$

The solution adopted to meet (6.40) and (6.41) is a spring and coil that act on the needle. To prevent the area from staying open while the system is disconnected, we use the coil to lift the needle and open the area. The spring will be responsible for closing it. If the mass of the needle is m_I, the rigidity of the spring is k_I, the lifting of the needle is h_I, and the initial precompression of the spring is h_0, the dynamic equation that describes the movement of the needle during closing is

$$m_I \frac{d^2 h_I}{dt^2} = -k_I (h_I + h_0) \tag{6.42}$$

To ensure quick closing, the force of the spring must be strong enough, which is achieved by also making the precompression and rigidity high enough. Under these conditions, $h_I \ll h_0$ can be verified. Consequently, equation (6.42) can be integrated to give the closing time

$$t_2 = \sqrt{\frac{2 m_I h_I}{k_I h_0}} \tag{6.43}$$

Condition (6.41) and equation (6.43) lead to the following design conditions:

$$m_I \rightarrow 0,\ h_I \rightarrow 0,\ k_I \rightarrow \infty,\ \text{y}\ h_0 \rightarrow \infty \tag{6.44}$$

The condition $h_I \rightarrow 0$ causes the tip of the needle (for which the shape is not known a priori) to interfere with the area of the outlet hole. For this reason, the condition $h_I \rightarrow 0$ is not viable, and must be replaced by the condition $h_I \gg A_f^{1/2}$. Moreover, to achieve quick opening, coil force $F(t)$ must be high enough. The dynamic equation governing the movement of the needle during opening is

$$m_I \frac{d^2 h_I}{dt^2} = -k_I h_0 + F(t) \tag{6.45}$$

Condition (6.40) requires that the force of the coil be much greater than the force of the spring

$$F(t) \gg k_I h_0 \tag{6.46}$$

Equation (6.46) makes it impossible to choose very high values of $k_I h_0$, which contradicts part of the result found in (6.44). The maximum force that the coil is capable of exerting will limit the value of $k_I h_0$. Nonetheless, condition (6.41) can be ensured by making the needle as small as possible ($m_I \rightarrow 0$) without increasing the production costs. This same condition also ensures that the dynamic response expressed by (6.45) will not prevent compliance with (6.40). However, (6.40) may be compromised because the force of the coil is a function of electric current i circulating through it. If L is the self-inductance of the coil, r is its electrical resistance, and v is its supply voltage, the differential equation governing the current in the coil is:

$$L\frac{di}{dt} + ri = v \tag{6.47}$$

This equation provides a characteristic response time for the current in the coil, during which the current changes by approximately its own value. It is therefore necessary to wait for a characteristic time, fixed by (6.47), for condition (6.46) to be met. This time is

$$t_1 \sim \frac{L}{r} \tag{6.50}$$

Condition (6.40) and equation (6.50) lead to

$$L \rightarrow 0\ y\ r \rightarrow \infty \tag{6.51}$$

Unlike mechanical time, which can be reduced as desired by decreasing the mass of the needle, electrical time cannot be reduced as desired. In fact, the condition $L \rightarrow 0$ cannot be carried out because when the self-inductance is reduced, the magnetic field decreases, along with the force of the coil. The same is true of the condition $r \rightarrow \infty$, which reduces the current and therefore the magnetic field and the force. Thus, time t_1 has a minimum value defined by the design of the coil, which cannot be reduced.[13]

Decisions adopted: Use a needle with a low mass. Use coils with the shortest response time possible.

Compliance with (6.36)

Condition (6.36) indicates that the fuel supply pressure must exceed the intake manifold pressure by the constant value ΔP, which must be set precisely. In order to meet this requirement, a pump must be placed between the fuel tank and the injector. A pump is a costly item, which is usually subcontracted to reduce costs. If the pump were responsible for supplying the pressure that verifies constraint (6.36), several theorems and corollaries would be violated.[14] The Minimum Tolerance Theorem is violated because the pump introduces new operation and design parameters that must be controlled by tolerances. In order for all of the pump's design parameters to be varied by roughly their own value, the system we design must be valid for any pump. If no tolerance is imposed on the pump, the pump is arbitrary.[15] Therefore, flow $\dot{m}_P$ and pressure P_P that the pump is capable of supplying must verify

$$\dot{m}_P \gg \dot{m}_f \text{ and } P_P \gg P_f \tag{6.52}$$

Conditions (6.52) do not verify condition (6.36), unless we add a device to the system to ensure compliance with (6.36). This system, as specified by the Minimum Tolerance Theorem, must have a single tolerance.[16] One possible solution is a pressure regulator comprised of a flat plate and a spring. The fuel pressurized by the pump is on one side of the flat plate, and air at the pressure of the intake manifold is on the other side. This flat plate maintains the pressure by means of the spring and a discharge hole, which is covered and uncovered by the flat plate. When the pump's supply pressure exceeds the value preset by the spring, the flat plate moves, opening the discharge hole (with a cross flow area A_D). Hence, pressure P_f drops until it reaches the value $P_a + \Delta P$. When the pump's supply pressure is lower than the value preset by the spring, the flat plate moves in the opposite direction, closing the discharge hole, and pressure P_f increases to $P_a + \Delta P$. The pressure regulator has a flat plate with mass m_P, area A_P, and position h_P, and a spring with rigidity k_P and precompression l_P. To avoid pressurized fuel leaks into the intake manifold, the flat plate must be sealed by a device exerting a force F_s dependent on the position and speed of the flat plate. Under these conditions, the dynamic equation for the flat plate is

$$m_P \frac{d^2 h_P}{dt^2} = A_P(P_f - P_a) - k_P(h_P + l_P) + F_s(h_P, \dot{h}_P) \qquad (6.53)$$

The number of tolerances in (6.53) can be reduced by decreasing the number of variables. Because the variable that appears the most is the flat plate position, it is advisable to restrict its movement ($h_P \rightarrow 0$).[17] This can be achieved if the discharge area is very large:

$$A_D \rightarrow \infty \qquad (6.54)$$

Equation (6.54) simplifies (6.53) and leads to the result fixed in (6.36):

$$\Delta P = P_f - P_a = \frac{k_P l_P}{A_P} \qquad (6.55)$$

The value of constant ΔP is chosen by fixing the values k_P, l_P and A_P. Therefore, to find the exact value of ΔP, we need to know the precise rigidity of the spring, its precompressed length, and the area of the flat plate. This means imposing more than three tolerances, which violates the Minimum Tolerance Theorem. To solve the problem, none of the parts of the pressure regulator may have tolerances.[18] The solution involves incorporating an adjustment device that makes it possible to vary the precompression of the spring. This device might be a simple screw, to be adjusted once the pressure regulator has been produced.[19]

Decisions adopted: The design operates $\forall$ pump which meets conditions (6.52). Constraint (6.36) is ensured by a pressure regulator, whose value is established by an adjustment process.

Fifth level

Condition $\forall_T \Delta t$ must be imposed by a train of electrical pulses with a width equal to Δt and voltage v. The pulse width must be set with the appropriate precision using a time generator. However, the pulse voltage is only used to generate a force in the coil that verifies (6.16). Therefore, ensuring that the voltage is over a minimum value, the system response is not very sensitive to its value.[20]

6.3.3 Analysis of the solution obtained

Once the conceptual design has been frozen, the detailed design must be created in order to apply it to a specific engine and check that the device verifies all of the constraints imposed by both the engine and the design process itself. The design obtained is the system now used on gasoline engines with multipoint electronic injection.[21] The advantages obtained due to compliance with the Theorems are:

1. Independence with regard to the engine. The solution obtained is capable of providing any fuel flow value by varying injection time Δt between a minimum value fixed by coil response time L/r, a maximum value fixed by the time it takes the crankshaft to rotate once $1/n$, and the response time of the coil. Because the engine's rotation regime is variable, the maximum opening time will be fixed by the maximum rotation regime n_{max} at which the engine will operate. The system designed will therefore be capable of operating for injection time values in the interval $[L/r, 1/n_{max} - L/r]$. Conditions (6.32) and (6.33) lead to $1/n_{max} - 2L/r \ll L/r$. Because the coil response times are in the order of a millisecond, the system designed will operate well for all gasoline engines with a maximum rotation regime well below 20,000 rpm.[22]
2. Independence with regard to the pump. Any sufficiently reliable pump available on the market is valid. This, in turn, means independence with regard to the pump supplier.
3. Independence with regard to the pressure regulator. Any pressure regulator that can be adjusted to the required value, and with the appropriate precision, is valid.
4. Independence with regard to the injector components. Only the area and discharge coefficient of the outlet hole on the injector must be accurately known. The needle, spring and coil have no tolerances critical for system behavior, and can therefore be designed to minimize production costs. The limitations are in the coil, which must provide the shortest response time possible, and the needle, which must have a low mass.
5. Independence with regard to the supply voltage. Within reasonable limits, the supply voltage does not affect the amount of fuel injected. Costly voltage stabilizers are not required. All of the power electronics required to switch the injectors can be provided with low-precision components.
6. Because there are only three tolerances affecting the system response (injector outlet area, pressure jump in the pressure regulator, and opening time), all other design parameters can be used to improve the system's application range.[23] Therefore, small variations in the

system make it possible to achieve other applications (direct injection of gasoline, etc.)

7. Independence with regard to production processes. Because there are only three critical tolerances, the rest of the dimensions (areas, volumes, finishes, etc.) can be specified to facilitate and reduce manufacturing and mass production costs.
8. Independence from noise (system robustness). Because there are only three critical tolerances, there are only three values for which variations can be problematic. For example, the system is insensitive to wear or leaks in the pump, changes in the pump rotation regime, fuel filter aging, etc.
9. The market entry barrier is high. Once the product has been placed on the market, it is difficult to create a new product that will improve performance, increase quality, or reduce costs. Economies of scale are the best strategy for competing on the market.

6.4 The principle of minimum generation of entropy and information

6.4.1 Gibbs' lemma and the evolution of systems subject to the laws of conservation

Gibbs' lemma, proven in Chapter 2, establishes that no set of non-negative variables whose sum is constant can have a higher entropy than that obtained by making all of the variables equal to each other. Given set a_i ($i = 1,2,\ldots,N$) of N variables, the only constraints imposed by the lemma are $a_i \geq 0$ and $\sum_{i=1}^{N} a_i = Na$, where a is a constant.[24] In particular, the lemma is always verified for dimensionless, non-negative variables, whose probability of occurrence is identical, and whose average is constant and equal to one. In other words, for the constraint

$$\frac{1}{N}\sum_{i=1}^{N}\frac{a_i}{a}=\sum_{i=1}^{N}\frac{a_i}{Na}=1 \tag{6.56}$$

Principle of conservation of energy: For an isolated system consisting of N free particles (interacting only through impacts between them) that do not exchange energy with the outside (the total energy is constant), it is verified that the maximum entropy is obtained when they all have the same kinetic energy in the three spatial directions, x, y and z. This constant enables us to define the temperature of the particles as follows (k is the Boltzmann constant):

$$\sum_{i=1}^{N} \frac{1}{2} m_i (\dot{x}_i^2 + \dot{y}_i^2 + \dot{z}_i^2) = E \tag{6.57}$$

$$\frac{1}{2} m_i \dot{x}_i^2 = \frac{1}{2} m_i \dot{y}_i^2 = \frac{1}{2} m_i \dot{z}_i^2 = \frac{E}{3N} = \frac{1}{2} kT \tag{6.58}$$

Principle of conservation of the linear momentum: For an isolated system comprised of N free particles (with no outside forces acting on it), the maximum entropy is obtained when half of the particles move in one direction, and the other half in the other (for each spatial direction). The absence of external forces is written as:

$$\sum_{i=1}^{N} m_i (\dot{x}_i \vec{u}_x + \dot{y}_i \vec{u}_y + \dot{z}_i \vec{u}_z) = 0 \tag{6.59}$$

If N_x are the particles moving towards the right in direction x, then (6.59) states that $N - N_x$ particles must move towards the left. Expression (6.60) is therefore true:

$$\sum_{i=1}^{N_x} m_i \dot{x}_i = \sum_{i=N_x+1}^{N-N_x} m_i \left|\dot{x}_i\right| \tag{6.60}$$

Each of the additions in the above expression must be equal to the other; therefore, the maximum entropy is obtained when all of the particles moving in the same direction have the same linear momentum. Let p_+ be the linear momentum of a particle moving towards the right, and p_- that of a particle moving towards the left. The maximum entropy condition enables us to write

$$\sum_{i=1}^{N_x} \frac{m_i \dot{x}_i}{N_x p_+} = \sum_{i=N_x+1}^{N-N_x} \frac{m_i \left|\dot{x}_i\right|}{(N - N_x)\left|p_-\right|} = \sum_{i=1}^{N_x} \frac{1}{N_x} = \sum_{i=N_x+1}^{N-N_x} \frac{1}{N - N_x} = 1 \tag{6.61}$$

In addition, the entropy of both sets is $\ln N_x$ and $\ln(N - N_x)$, respectively. The total entropy is therefore $\ln N_x + \ln(N - N_x)$, which reaches the maximum for the condition $N_x = N/2$. This verifies $p_+ + p_- = 0$ and

$$m_i|\dot{x}_i| = m_i|\dot{y}_i| = m_i|\dot{z}_i| = p_+ \tag{6.62}$$

Principle of conservation of mass: For an isolated system comprised of N free particles distributed in homogeneous volume V, divided into n identical cells (with no outside forces or other elements to distinguish any cells from the others), it is verified that the maximum entropy is obtained when each cell contains the same number of particles.

$$\sum_{i=1}^{N} m_i = Nm = \sum_{j=1}^{N_1} m_i + \sum_{j=N_1+1}^{N_2} m_i + \ldots + \sum_{j=1}^{N} m_i \tag{6.63}$$

$$\sum_{j=1}^{N_1} m_i = \sum_{j=N_1+1}^{N_2} m_i = \ldots = \sum_{j=1}^{N} m_i = \frac{N}{n} m \tag{6.64}$$

Ideal gas law: A set of N particles enclosed in volume V, which verifies the three laws of conservation above and has reached the state of maximum entropy, must verify the condition $PV = NkT$, where P is the pressure exerted on the walls of the volume. Indeed, if a wall is placed perpendicular to direction x, the particles traveling towards it that are located at distance Δx will reach the wall in time $\Delta t_i = \Delta x/\dot{x}_i$. If all particles are identical ($m_i = m$, $\Delta t_i = \Delta t$), the number of particles traveling towards the wall with area A is $N_x \Delta x A/V$. The variation of their linear momentum is $2p_+ N_x \Delta x A/V$. This variation of the linear momentum is produced by force F exerted by the wall during time Δt. Consequently, $p_+ N \Delta x A/V = F \Delta t$ is verified. Thanks to (6.62), we can therefore conclude

$$2N\frac{1}{2}m\dot{x}^2 = \frac{F}{A}V \tag{6.65}$$

Remembering (6.58), we find the relationship:

$$NkT = PV \tag{6.66}$$

This equation has been experimentally corroborated. It is therefore a reality that nature evolves, in those systems meeting the conditions in Gibbs' lemma, searching for the maximum entropy; i.e. searching for uniform distributions of temperature, pressure, concentrations and probability distributions.[25]

6.4.2 The Central Limit Theorem

Expressions (A.3) and (A.11), explained in the Appendix, establish that variable z, obtained by adding $N - 1$ independent variables y_i with mean η and standard deviation σ, has a normal probability density function:

$$z = \frac{\sum_{j=1}^{N-1} y_i - (N-1)\eta}{\sqrt{N-1}\sigma} \xrightarrow[N\to\infty]{} \frac{e^{-\frac{z^2}{2}}}{\sqrt{2\pi}} \tag{6.66}$$

This expression determines that the standard deviation of a sum of independent variables grows as the root of the number of variables added. Equation (6.66) enables us to evaluate the variation of one of the variables appearing in the constraint for Gibbs' Lemma

$$\frac{1}{N}\sum_{j=1}^{N} y_j = y = 1 \tag{6.67}$$

$$z = \frac{N(y-\eta)+\eta-y_N}{\sqrt{N-1}\sigma} \xrightarrow[N\to\infty]{} \frac{e^{-\frac{z^2}{2}}}{\sqrt{2\pi}} \tag{6.68}$$

Because the mean of variable y_N is also η, $\eta = y = 1$ must be verified. Thus, we find that the closing variable that ensures exact compliance with the condition required by Gibbs' Lemma is distributed normally, and has a standard deviation that grows as $N^{1/2}$.

$$\frac{1-y_N}{N^{1/2}\sigma} \xrightarrow[N\to\infty]{} \frac{e^{-\frac{z^2}{2}}}{\sqrt{2\pi}} \tag{6.69}$$

From (6.69), we conclude that $y_N \in [1 - 2N^{1/2}\sigma, 1+2N^{1/2}\sigma]$ holds 96 per cent of the occurrences (see Appendix). Because the variable y_N must be non-negative, we find that

$$1 - 2N^{1/2}\sigma \geq 0 \tag{6.70}$$

Hence,

$$\frac{1}{2N^{1/2}} \geq \sigma \tag{6.71}$$

Expression (6.71) indicates that the variables fluctuate with a lower standard deviation, the higher the number of variables considered by Gibbs' Lemma. Thus, the higher the number of variables, the lower the oscillation of the variables around the uniform solution. Nonetheless, for a finite number of variables, the fluctuation, although slight, always exists. Therefore, the Central Limit Theorem and Gibbs' constraint ensure that when there is a sufficiently large number of interchangeable variables, the distribution that they tend towards is a uniform distribution with an error that decreases as the root of the number of variables.[26]

6.4.3 Design as a process for reducing entropy and information

In Chapter 2, design was defined as the process that decreases the entropy of the alphabets defining the possible solutions. We modified their distributions by comparing the Space of Definition for the Needs to the Space of Definition for the Response, and specified the result in the Space of Definition for the Satisfaction (where the relevant information was the probability associated with the 'acceptance' label). The uncertainty associated with the joint probability of these 'acceptance' labels was the information content (3.29) used to formulate the Information Axiom.[27] However, the decrease in the average uncertainty of the rest of the alphabets (for example, the alphabets defining the solution) must be calculated using expression (2.19) or (2.30). When a designer has no arguments (lacks information) to establish the best labels for X, the probability distribution associated with X is the uniform distribution. When the designer has arguments for ruling out certain labels and keeping others,[28] he introduces the amount of information given by (2.19). However, even if the designer introduces the greatest amount of information possible (given by equation (2.22)), the solution selected may not be the best one, and might not even be valid. This issue is resolved by the minimum uncertainty in the 'acceptance' or 'success' labels. The information in (2.19) and (2.30) therefore only measures information at the syntactic level (only considering the possible evolution of the signs without understanding their meaning), but not the pragmatic level (see Section 2.8). To ensure that the entropy reduction process during the design process selects a label with a maximum probability of success, the designer must be perfectly familiar with the meaning of the labels and the argumentations relating them (semantic level), and communicate this information when required (pragmatic level). For this reason, mutual information and information content are independent concepts.

Starting from a uniform distribution, the designer can only reduce the entropy of all of the spaces (according to property (2.24), he can only add information). The Information Axiom also states that reducing the entropy in the Space of Definition for the Satisfaction will cause the label with the greatest probability to be the 'acceptance' label. We can therefore define design as:

Definition: The *design process* is the activity that formulates a plan to systematically reduce entropy until achieving the satisfaction of a need with the maximum probability of success.

Because reducing entropy goes against the natural tendency to increase it (described briefly in Sections 4.1 and 4.2), there is a consumption of resources that must be minimized.[29] If this is accomplished, the design process will be efficient. In expression (2.30), we can see that there are two terms capable of increasing the information, and therefore the resources consumed: 1) initial entropy, and 2) final entropy.

The *principle of minimum generation of entropy and information* establishes that the initial entropy must be minimal, and the final entropy must be as close to the initial entropy as possible.[30]

The Broad Tolerance Theorem tends to increase the width of the intervals where the different variables intervening in the design can take values. Therefore, according to the definition of differential entropy (Eq. (3.38)), the greater the width of the interval, the greater the differential entropy associated with that variable, and the greater the final entropy. Relaxing tolerances is therefore beneficial because it decreases the cost of fighting against the natural tendency of variables to increase their entropy. In addition to this benefit, the use of broad tolerances reduces the upper extreme of the entropy of the spaces generated during the design process (see Statement #4). It also reduces the information needed for the design (Statement #1) and the number of design parameters, as a parameter with very broad tolerances can adopt any value without affecting the response (Corollary 2*). Consequently, its value is indifferent for the design.

Moreover, the design methodology described in this book tends to reduce the number of functional requirements and design parameters; in other words, the number of variables appearing in the spaces of definition for the needs and the solution. As we saw in Chapter 2, the smaller the number of variables, the lower the upper extreme of the entropy of the spaces created (see Statements #2 and #3). These measures reduce the final entropy.

Entropy-Based Design compiles the above results into four rules that constitute a guide for the designer. The result is the procedure shown in the example in Section 6.3, which can be summarized as follows.

Basic Steps of Entropy-Based Design: On each level of the design hierarchy, follow these four rules:

1. Resolve only one functional requirement at a time. (If this is not possible, use the Independence Axiom).
2. To meet the functional requirement, select the design parameter that provides the most linear response (Linearity Theorem), and that introduces the minimum number of new parameters (Minimum Tolerance Theorem).
3. Use the broadest tolerances possible (Broad Tolerance Theorem).
4. Maximize the probability of success (Information Axiom).

Conclusion: The Advanced Design Theories (Axiomatic Design, Metric Design, Reliability-Based Design and Entropy-Based Design) drastically reduce the entropy and information involved in all design activities. They also provide a very universal definition of what 'best' means in engineering, for both the product and the design process.[31]

6.5 Notes

1. The needs $y_1 = E[y]$ and $y_2 = E[(y - E[y])^2]$ are independent because it is always possible to vary the position of an interval without modifying its width, and vice-versa. Therefore, according to the definition given in Chapter 3, both needs form a set of functional requirements.
2. The author does not know of any reference where this theorem is stated and proven.
3. This example specifies a single functional requirement on each level of the design hierarchy, and therefore automatically verifies the Independence Axiom.
4. Subscript T indicates that at some point in the design process, that value will be fixed with the appropriate precision.
5. We have made ourselves independent from the customer who hired us as designers. Thus, we have also turned everyone else into possible customers. In other words, we have improved our chances of accessing a larger market, and thereby increased our future market share. This benefit comes from applying the methodology described in this book.
6. The objective of the advanced design methodologies is to directly freeze the best solution to the problem. The purpose is to take only that solution to the detailed design phase. This reduces the costs associated with the design process.
7. An expert on piston engines could choose to pass through the cylinder head and inject the fuel directly in the chamber. The solution that the proposed methodology will provide is so versatile that, with slight modifications, it will be able to satisfy other configurations.
8. If it is determined in a later stage of the design hierarchy that the choice of area requires more than one tolerance, this decision should be questioned.
9. Practically none of the variable-area systems that we can imagine for satisfying $\forall_T A_f$ meets the Theorem of the Minimum Number of Tolerances or the Linearity Theorem. Remember that we must create a system with variable geometry (with a linear behavior and a single tolerance) capable of providing any passage area value from a minimum (for example, zero) to a maximum. Consider whether a ball or gate valve, a clamp acting on a flexible tube, or a conical needle entering a hole (or any other type of valve) would meet both theorems.
10. These equations are design equations because they are imposed by the design methodology used. If the design methodology were to change, equations (6.32) and (6.33) might be different or unnecessary.
11. If n is the engine's rotation regime, it is not a selectable design parameter. However, if n is chosen as an independent frequency of the engine's rotation regime, it could be used for flow control. Because it is linear, $\forall_T n$ could be chosen in order to find $\forall_T \dot{m}_f$. It is an interesting exercise, left to the reader, to explore the result that would be obtained in this case.

12. Nowhere is it specified that stationary processes are better than non-stationary processes. In this case, compliance with the Linearity Theorem has led the design towards a discontinuous fuel supply system, independently of the type of engine used (continuous or discontinuous). However, the same theorem also required the system response to be quasi-steady within each time interval Δt.
13. This time sets both the closing and opening values because the characteristic time in which the electric current increases is the same time it takes for it to decrease. Another possible solution is to use a coil that does not have to meet condition (6.46). In this case, a hydraulic system must be installed to amplify the force of the coil until (6.46) is met. This is the solution adopted in the common-rail injectors for Diesel engines (IMechE, 2009).
14. Corollary (2*) is violated because all of the operation and design parameters for the pump are incorporated into the design. The Metric Design Tolerancing Theorem is violated because the pump is a costly item.
15. The pumps will be chosen so that (6.52) is verified, and must be cheap and reliable.
16. As specified by the Metric Design Tolerancing Theorem, it should also be cheaper than the pump.
17. This condition cannot be verified if the engine frequency excites the frequency of the pressure regulator itself. Nonetheless, if such a situation occurs, it is easy to damp the pressure oscillations by adding expansion tanks to the system (which increase in volume if the pressure rises, and decrease if it drops) and dissipators. Reducing the displacement of the flat plate also allows it to be built into the regulator case. This will ensure fueltightness. It also enables it to be produced by stamping, drastically reducing production costs.
18. Obviously, the production process for the pressure regulator might require the use of tolerances; however, none of those tolerances will affect the behavior of the system we are designing (as long as they allow for compliance with reliability and cost constraints). For this reason, the pressure regulator production process is completely uncoupled from the design of the fuel supply system. All degrees of freedom can be used so that the production engineers achieve a pressure regulator with the minimum cost and maximum reliability. This independence of the regulator from the engine promotes the benefits derived from an economy of scale.
19. The Metric Design Tolerancing Theorem is met because the pressure regulator has a much lower cost than the price of the pump.
20. Although this is true for the opening time, it might not be for the closing time. Closing occurs when the current drops below the value that keeps the needle stuck to its upper stop. Very high voltage values may cause very high current values, which can cause the needle to stick to the upper stop for a long time.
21. Unlike direct injection systems (which inject into the chamber), these systems supply fuel close to the corresponding intake valve.
22. Unless modifications are made to the design, Formula 1 engines will not be able to use the system directly. Certain modifications would be required to use this system on an F1 car. It is an interesting exercise, left to the reader, to

discover how to modify the system (without adding new tolerances) in order to improve its response time.

23. For example, two voltage levels can be used to supply the coil: the first much higher than the supply voltage in order to force quick opening of the injector, and the second close to the level that prevents the injector from closing.
24. Remember the theorem proven in Chapter 2, which establishes that uniform distribution (for discrete variables) has maximum entropy:

$$-\sum_{i=1}^{N} a_i \log a_i \leq -\left(\sum_{i=1}^{N} a_i\right)\log\frac{\sum_{i=1}^{N} a_i}{N} = -Na\log a.$$

25. Uniform distribution is a limit towards which nature tends. Nonetheless, reality approaches it as much as possible, but never reaches it exactly due to the fluctuations existing in all of the variables. For example, the speed of all particles is not exactly the constant value $\dot{x}_i = p_+/m_i$, but suffers variations due to different noise sources (for example, the collisions between particles). Thus, each particle is affected by the random component δ_i, in such a way that the following is met: $m_i\dot{x}_i = p_+ + \delta_i \neq p_+ + \delta_j = m_j\dot{x}_j$.
26. If equation (6.67) were replaced by $\sum_{j=1}^{N} y_j = 1$, the fluctuation with respect to the uniform distribution would have decreased as $N^{-3/2}$.
27. Suh (1990: 151) defines the information content as the logarithm for the inverse of probability $I = -\log_2 p$ (see Eq. (3.29)). This is known as the uncertainty (Eq. (2.1)). According to Nam P. Suh's definition, the information content is the measure of the knowledge needed to satisfy a functional requirement on a particular level of the design hierarchy (Suh 1990: 65). The greater the probability of success desired for the resolution of the requirement, the greater the knowledge that must be available to reduce the unforeseen circumstances. If success is understood as the satisfaction of the functional requirements, then the information is the measure of the knowledge required to satisfy a given set of functional requirements on a particular level of the design hierarchy (65). It was later acknowledged (154) that when there is a discrete number of events occurring in a set whose probabilities are $p_1, \ldots, p_i, \ldots$, the above definition adopted for the information content must be extended to calculate the average information content of the discrete events as $I = -\sum_i p_i \log p_i$. This is the definition of entropy (see Eqs. (2.1) and (2.2)). The difference between the maximum entropy principle and the minimum information principle is that the former seeks the probability distribution that maximizes the entropy subjected to a series of constraints, and the latter seeks the distribution that minimizes it subjected to another set of constraints (Suh, 1990: 155).
28. For example, Equation (2.40) or any of the theorems explained in the book can be used as arguments.
29. The only source of information destruction is the natural evolution described in Sections 4.1 and 4.2.
30. Remember that there are two types of operators in the design process (see Section 2.12). One of the operators performs entropy-generating activities, and the other performs entropy-reducing activities. The first generates

entropy by creating the spaces of definition. The second reduces entropy by decreasing the uncertainty of the probability distributions associated with the spaces of definition.

31. The postulate accepted throughout the book is that the best design is the one obtained by applying these methodologies. This Section summarizes them by establishing that the best design must accomplish the Principle of Minimum Generation of Entropy and Information. The four Entropy-Based-Design rules presented here aim at this purpose.

Appendix: statistical concepts

The numbers used by a designer (values used as data or results) are normally affected by noise and inaccuracies that turn them into random variables. Thus, the values taken by the variable y at a particular moment form a set $\{y_1,...,y_N\}$, which we will refer to as an N-sized sample. These values also belong to a larger set, called a population, whose elements are all the possible results that could appear in the sample.

The elements of the population have a particular possibility of appearing in a sample. We use the term *probability distribution* to refer to the function that assigns each element of the population a probability of appearing in an $N = 1$-sized sample. The mean (population mean) and variance (population variance) of the population are:

$$\eta = E[y] \tag{A.1}$$

$$\sigma^2 = V[y] = E\left[(y-\eta)^2\right] \tag{A.2}$$

The researcher does not know the values defined in (A.1) and (A.2), but can calculate the mean (sample average) and variance (sample variance) of the sample:

$$\hat{\eta} = \frac{\sum_{j=1}^{N} y_j}{N} \tag{A.3}$$

$$\hat{\sigma}^2 = \frac{\sum_{j=1}^{N} \left(y_j - \hat{\eta}\right)^2}{N-1} \tag{A.4}$$

The values $\hat{\eta}$ and $\hat{\sigma}$ are also random variables whose probability distributions have the following means and variances:

$$E[\hat{\eta}] = \eta \tag{A.5}$$

$$V[\hat{\eta}] = \frac{\sigma^2}{N} \tag{A.6}$$

$$E\left[\hat{\sigma}^2\right] = \sigma^2 \tag{A.7}$$

Equation (A.6) establishes that if N is very large, the variance of $\hat{\eta}$ tends to zero. However, the addition of variables increases the variance. Indeed, the probability distribution associated with the random variable obtained by adding or subtracting two independent random variables has the following mean and variance:[1]

$$E\left[y_A \pm y_B\right] = \eta_A \pm \eta_B \tag{A.8}$$

$$V\left[y_A \pm y_B\right] = \sigma_A^2 + \sigma_B^2 \tag{A.9}$$

A.1 Central limit theorem

Let y be a random variable that takes values from a population whose probability distribution function has a well-defined mean η and variance σ^2. Let $\{y_1, \ldots, y_N\}$ be a sample containing N elements obtained independently from that population. The theorem establishes that when the sample size tends to infinity, the mean of the sample defined according to (A.3) meets

$$\lim_{N\to\infty} \Pr\left[\frac{\hat{\eta}-\eta}{\sigma/\sqrt{N}} \le z\right] = \frac{1}{\sqrt{2\pi}} \int_{-\infty}^{z} e^{-\frac{\xi^2}{2}}\, d\xi = \Pr_{N(0,1)}\left[(-\infty, z)\right] = \Pr_{N(0,1)}\left[z\right] \tag{A.10}$$

Therefore, if the sample size is large enough, the associated probability density function is

$$\lim_{\substack{N\to\infty \\ dz\to 0}} \frac{\Pr\left[z < \dfrac{\hat{\eta}-\eta}{\sigma/\sqrt{N}} \le z+dz\right]}{dz} = \lim_{dz\to 0} \frac{\Pr_{N(0,1)}\left[z+dz\right] - \Pr_{N(0,1)}\left[z\right]}{dz} = \frac{e^{-\frac{z^2}{2}}}{\sqrt{2\pi}} \tag{A.11}$$

A.2 Normal distribution

The central limit theorem (A.11) defines a probability distribution with mean 0, standard deviation 1, and probability density function

$$pdf(z) = \frac{e^{-\frac{z^2}{2}}}{\sqrt{2\pi}} \tag{A.12}$$

This distribution is known as a normal distribution (Gaussian distribution), and is represented in abbreviated form as $N(0,1)$.

A.3 Sample of a population with a normal distribution

If y is a random variable whose probability distribution function is normal with mean η and variance σ^2 (abbreviated to $N(\eta, \sigma)$), then the random variable $(y - \eta)/\sigma$ has the normal distribution function $N(0,1)$ associated with it. If the variable $\hat{\sigma}^2$ has been calculated, as shown in (A.4), based on N independent observations obtained from the normal population $N(\eta, \sigma)$, then the random variable

$$z = \frac{y-\eta}{\hat{\sigma}} \tag{A.13}$$

has a distribution function known as Student's distribution, with $N - 1$ degrees of freedom (Box et al., 1978). If the number of elements in the sample used to calculate $\hat{\sigma}$ were infinite, then $\hat{\sigma}$ would coincide with σ, and Student's distribution with infinite degrees of freedom would coincide with $N(0,1)$. The lower the number of degrees of freedom used to calculate $\hat{\sigma}$, the flatter the Student's distribution curve will appear, and the larger the area enclosed by its tails. In particular, if we take N observations of a variable distributed according to $N(\eta, \sigma)$, then we can state:

1. The distribution of $\hat{\eta}$ is distributed according to $N(\eta, \sigma / \sqrt{N})$.
2. The statistical $(N - 1)\hat{\sigma}^2/\sigma^2$ is distributed independently of $\hat{\eta}$ with the distribution χ^2 and $N - 1$ degrees of freedom.
3. The statistical $\dfrac{\hat{\eta}-\eta}{\hat{\sigma} / \sqrt{N}}$ is distributed with $N - 1$ degrees of freedom in the Student's distribution. If N tends to infinity, the distribution is $N(0,1)$.

These distributions can be used to find confidence limits for the estimated parameters. The confidence interval $[-z_{\alpha/2}, z_{\alpha/2}]$ is the one whose values do not produce a significant discrepancy, defined by α, with the actual data.[2] The degree of discrepancy is given by probability α that the actual data is outside the confidence interval. For example, the confidence interval for the mean of a population from which a sufficiently large N-sized sample has been taken is given by $\eta \in \left[\hat{\eta} - z_{\alpha/2}\hat{\sigma} / \sqrt{N}, \hat{\eta} + z_{\alpha/2}\hat{\sigma} / \sqrt{N}\right]$, where $z_{\alpha/2}$ is the value that leaves an area equal to $\alpha/2$ under the upper tail of distribution $N(0,1)$. Thus, confidence interval $\left[\hat{\eta} - z_{\alpha/2}\hat{\sigma}/\sqrt{N}, \hat{\eta} + z_{\alpha/2}\hat{\sigma}/\sqrt{N}\right]$ is an estimation of the value of η with a probability given by the confidence coefficient $1 - \alpha$.

A.4 Component lifetime

The lifetime of a component with respect to a particular failure mode can be considered a random variable. Similarly, for preset time t, the variable $\theta_j(t)$, which is equal to 1 if device j has failed in that failure mode before that time, and equal to 0 if it has not, is a random variable. If a sample of N devices is taken, and they are tested up to preset time t, the number of devices that have failed (in that failure mode) will be $N_F(t) = \sum_{j=1}^{N} \theta_j(t)$.

Therefore, the probability of occurrence for that failure mode can be estimated as $\hat{F}(t) = N_F(t)/N$. In other words, the estimated probability of failure is the mean for sample $\{\theta_1(t), \ldots, \theta_N(t)\}$, which in turn is a random variable whose mean $E[\theta(t)]$ is the true probability of failure $F(t)$. Consequently, based on $\theta_j(t)$, we can define the following random variables

$$\hat{F}(t) = \frac{\sum_{j=1}^{N} \theta_j(t)}{N} \tag{A.14}$$

$$\hat{\sigma}(t)^2 = \frac{\sum_{j=1}^{N} \left(\theta_j(t) - \hat{F}(t)\right)^2}{N-1} = \frac{\sum_{j=1}^{N} \left(\theta_j(t) - \hat{F}(t)^2\right)}{N-1} = \frac{N}{N-1} \hat{F}(t)\left(1 - \hat{F}(t)\right) \tag{A.15}$$

The above expression includes random variable $\hat{F}(t)^2$, whose mathematical expectation can be calculated in the event that each specimen in the sample evolves independently from the others:

$$E\left[\hat{F}(t)^2\right] = \frac{\sum_{i=1}^{N} \sum_{\substack{j=1 \\ j \neq i}}^{N} E\left[\theta_i(t)\right] E\left[\theta_j(t)\right] + \sum_{i=1}^{N} E\left[\theta_i(t)\right]}{N^2} = \frac{(N-1)F(t)^2 + F(t)}{N} \tag{A.16}$$

Taking this expression into account, it is easy to prove that

$$E[\hat{F}(t)] = E[\theta(t)] = F(t) \tag{A.17}$$

$$E\left[\hat{\sigma}(t)^2\right] = \frac{N}{N-1}\left(E\left[\hat{F}(t)\right] - E\left[\hat{F}(t)^2\right]\right) = F(t)\left(1 - F(t)\right) \tag{A.18}$$

Equations (A.5), (A.6), (A.7), (A.17) and (A.18) enable us to define the variable

$$z(t) = \frac{\hat{F}(t) - F(t)}{\sqrt{\frac{F(t)\left(1 - F(t)\right)}{N}}} \tag{A.19}$$

The central limit theorem shows that if the number of elements in the sample is large enough, the random variable $z(t)$ is distributed by $N(0,1)$. For a confidence coefficient of 96 per cent ($\alpha = 0.04$, $z_{0.02} \simeq 2$), the expected probability of failure for the sample is

$$\hat{F}(t) = F(t)\left[1 \pm 2\sqrt{\frac{1 - F(t)}{NF(t)}}\right] \tag{A.20}$$

The absolute error committed by the estimation will be:

$$\varepsilon = \hat{F}(t) - F(t) = \pm 2\sqrt{\frac{F(t)\left(1 - F(t)\right)}{N-1}} \tag{A.21}$$

The product $(1 - F(t))F(t)$ is maximum when $F(t) = 1/2$. This enables us to set the following bound (valid for large values of N)

$$|\varepsilon| \leq \sqrt{\frac{1}{N-1}} \ll 1 \tag{A.22}$$

Therefore, if the sample size increases, the error committed when replacing $F(t)$ with the estimation $\hat{F}(t)$ tends to zero as $N^{-1/2}$, i.e. $F(t) \simeq \hat{F}(t) + 0(N^{-1/2})$ is met. This result, along with (A.20), provides

$$\hat{F}(t) = F(t)\left[1 \pm 2\sqrt{\frac{1 - \hat{F}(t) - 0(N^{-1/2}))}{N\left(\hat{F}(t) + 0(N^{-1/2})\right)}}\right] \tag{A.23}$$

Isolating the mean for the population in (A.23), we discover an expression that bounds the mean with a confidence level of 96 per cent.

$$F(t) = \hat{F}(t)\left[1 \mp 2\sqrt{\frac{1 - \hat{F}(t)}{N\hat{F}(t)}} + 0\left(\frac{1}{N}\right)\right] \tag{A.24}$$

The second term inside the brackets is made arbitrarily large if $\hat{F}(t)$ is close to zero. In practice, this forces us to take a larger sample the lower the probability of failure, i.e. $N \gg 1/\hat{F}(t)$ must be verified. When the sample size tends to infinity, expression (A.24) defines function $F(t)$ with a null error. By following the above process for different time values, we obtain an approximation of the probability of failure as a function of time. Based on this approximation, the reliability is

$$R(t) = 1 - F(t) \tag{A.25}$$

Expression (A.24) shows that if N is large enough, the variable $z(t)$ defined in (A.19) (whose distribution is $N(0, 1)$ if N tends to infinity) can be approximated by

$$z(t) \simeq \frac{\hat{F}(t) - F(t)}{\sqrt{\dfrac{\hat{F}(t)\left(1-\hat{F}(t)\right)}{N}}} \tag{A.26}$$

A.5 Number of failed parts in an infinitesimal time interval

If we take an N -sized sample, the number of failures that will occur in the sample before time t will be $N_F(t) = N\hat{F}(t)$, which is a random variable. Keeping in mind that when N tends to infinity, the variable defined in (A.19) follows the distribution $N(0, 1)$, we can establish that for a confidence coefficient of 96 per cent, the expected probability of failure for a sample is bound by

$$N\hat{F}(t) = NF(t)\left[1 \pm 2\sqrt{\frac{1-F(t)}{NF(t)}}\right] \tag{A.27}$$

This expression states that for a sufficiently large sample, the number of parts that fail before time t is $F(t)N$. Thus, the number of failures between t and $t + dt$ is

$$F(t+dt)N - F(t)N = \frac{dF(t)}{dt}Ndt + 0(dt^2) \tag{A.28}$$

Therefore, the number of parts that fail is proportional to the probability density function

$$pdf(t) = \frac{dF(t)}{dt} \tag{A.29}$$

A.6 Definition of lifetime

As we have seen, $F(t)$ is the probability that a failure will occur in a component before time t. We will refer to the lifetime of device j as the instant t_j in which it fails. As discussed, time t_j is a random variable. The probability that this time will be less than t is given by the probability that it will fail before t. In other words,

$$\Pr[t_j < t] = F(t) = \Pr[\theta_j(t) = 1] \tag{A.30}$$

The probability that it will fall within the interval $[t_A, t_B]$ is

$$\Pr[t_A \le t_j < t_B] = F(t_B) - F(t_A) \tag{A.31}$$

The probability that the instant of failure will fall within the interval $[t, t + dt]$ is

$$\Pr[t \le t_j < t + dt] = F(t + dt) - F(t) = \frac{dF(t)}{dt} dt + 0(dt^2) = pdf(t)dt + 0(dt^2) \tag{A.32}$$

A.7 Probability density function for the number of failures and lifetime

In the previous sections, we have seen that the number of failures expected before instant t is distributed normally around a mean value defined by $NF(t)$. According to (A.27), the deviation of the number of failures with respect to $NF(t)$ is below $\pm 2\sqrt{NF(t)(1 - F(t))}$ 96 per cent of the time. Moreover, the lifetime of a specimen is described by the probability density function $pdf(t) = dF(t)/dt$ given in (A.29). A single known probability density function is sufficient for calculating the expected values based on both random variables. This density function will depend on the device, the failure mode, and the operating conditions. A statistical model is a density function with one or more unknown constants that are experimentally determined. Some models are discussed below, along with a method for estimating their constants.

A.7.1 Hazard rate

It is useful to define the hazard rate as

$$h(t)=\lim_{\Delta t\to 0}\frac{R(t)-R(t+\Delta t)}{R(t)\Delta t}=-\frac{1}{R(t)}\frac{dR(t)}{dt}=\frac{f(t)}{R(t)}=-\frac{d}{dt}Ln\,R(t) \qquad (A.33)$$

This rate corresponds to a process where the specimens that have failed are replaced by other specimens identical to the failed ones in terms of remaining life, rather than new specimens (Tobias and Trindade, 1986). From the above expression, we see that the reliability can be written, without loss of generality, as the following exponential

$$R(t)=e^{-\int_0^t h(y)\,dy} \qquad (A.34)$$

A.7.2 Mean time between failures (MTBF)

A component failure is usually resolved by replacing the failed specimen with a new one that will also fail after some time has passed. During a period of continuous system use, N failures and N replacements will have occurred. The total system operation time will be $t_1 + t_2 + \ldots + t_N$, and an estimation of the mean time between failures will be

$$\hat{t}=\frac{\sum_{j=1}^{N} t_j}{N} \qquad (A.35)$$

According to (A.5) and (A.6), the mean value shown above has the following mathematical expectation and variance

$$E[\hat{t}]=E[t]=\int_0^\infty t\,pdf(t)\,dt \qquad (A.36)$$

$$V[\hat{t}]=\frac{V[t]}{N}=\frac{1}{N}\int_0^\infty (t-E[t])^2\,pdf(t)\,dt \qquad (A.37)$$

When the number of replacements performed is very high, the variance tends to zero and the mean time between failures tends to

$$MTBF=\hat{t}=\int_0^\infty t\,pdf(t)\,dt=-\int_0^\infty t\,dR=-tR\Big|_0^\infty+\int_0^\infty R\,dt=\int_0^\infty R(t)\,dt \qquad (A.38)$$

A.7.3 Failure rate

The average frequency with which failures occur is the inverse of the mean time between failures.

$$\lambda = \frac{1}{MTBF} \tag{A.39}$$

A.7.4 Exponential model

Observation of expressions (A.34) and (A.39) enables us to generate an approximate model where the MTBF is constant.

$$R(t) \approx e^{-\lambda t} \tag{A.40}$$

A.7.5 Weibull model

This is a generalization of the exponential model which assumes that specimen aging is such that an increase in aging (adimensionalized with the cumulative aging) is proportional to the time increment during which it occurs (adimensionalized with the elapsed time). The proportionality constant is damage exponent s, and variable t^s is a measure of aging.

$$R(t) \approx e^{-\lambda t^s} \tag{A.41}$$

Tobias and Trindade (1986) and Nelson (2004) provide variations of the Weibull model based on acceleration factors. When the acceleration factor follows a potential law, this is a particular case of the model for variable loads, which will be presented below.

A.7.6 Weibull model for variable loads

This is used to calculate the lifetime of specimens subjected to a load $Q(t)$ that varies over time. Benavides (2010) presents a justification of this model, and an application of damage exponent calculations in thermal shock and thermal fatigue tests.[3]

$$R(t) \approx e^{-\left(t\frac{Q_e(t)^p}{k^p}\right)^s} \tag{A.42}$$

Where k, s and p are constants and $Q_e(t)$ is an equivalent load calculated using the expression

$$Q_e(t) = \left[\frac{s}{t^s}\int_0^t Q(\tau)^{ps}\tau^{s-1}\,d\tau\right]^{\frac{1}{ps}} \tag{A.43}$$

For conventional products, such as general-purpose industrial bearings, lifetime is measured in millions of cycles, $t = 10^6\ L/\upsilon$, where υ is the frequency with which the load cycles occur. In addition, the constant k is chosen so that the lifetime of a device subjected to a constant load with a 10 per cent probability of failure follows a potential law equal to $L_{(F = 0.1)} = (C/Q)^p$. This causes constant C, the dynamic load capacity, to appear in the above expression. Thus, depending on the units chosen for lifetime and the value set for reliability, the constant that adimensionalizes the load takes different values.

$$R(t) = e^{-s\int_0^t \left(\frac{Q(\tau)}{k}\right)^{ps} \tau^{s-1}\, d\tau} \tag{A.44}$$

$$R(L) = e^{\ln(0.9)s\int_0^L \left(\frac{Q(L)}{C}\right)^{ps} L^{s-1}\, dL} \tag{A.45}$$

A.7.7 Maximum likelihood estimator

In practice, in an instant of time, multiple specimens are being operated or tested, or have already been operated and tested. Each one will have been operated during time t_j, and will have been subjected during that time to an equivalent load given by $Q_{ej}(t_j)$, obtained by integrating the variable load $Q_j(t)$ imposed on the specimen, as shown in (A.43). (Equations (A.25) and (A.42) provide $F(t)$, and (A.29) provides $pdf(t)$.) During that time, the following events can occur:

1. The specimen failed before time t_j, but the failure was not detected until instant t_j. The probability of this event is given by $F(t_j)$. We will assume that there are N_1 specimens in this situation.
2. The specimen failed before time t_j and after time $t_j - T_j$. This event occurs when specimen failure is checked at preset time intervals. Therefore, if a failure is observed, it is only known that it occurred after the second to last inspection and before the last one. The probability of this event is given by $F(t_j) - F(t_j - T_j)$. We will assume that there are N_2 specimens in this situation.
3. The specimen failed exactly at instant t_j. The probability of this event is $pdf(t_j)dt$. We will assume that there are N_3 specimens in this situation.
4. The specimen did not fail before time t_j. This event occurs: 1) if the specimen failed at instant t_j, but due to a different failure mode than the one under study, 2) if operation or testing had to be suspended for any other reason, or 3) If it continues to work. Consequently, if

the failure mode under study did not occur during time t_j, the probability of this event is $1 - F(t_j) = R(t_j)$. We will assume that there are N_4 specimens in this situation.

The probability of this situation occurring is proportional to

$$\Omega = \prod_{i=1}^{N_1} F(t_i) \prod_{j=N_1+1}^{N_1+N_2} \left(F(t_j) - F(t_j - T_j)\right) \prod_{k=N_1+N_2+1}^{N_1+N_2+N_3} pdf(t_k) \prod_{l=N_1+N_2+N_3+1}^{N_1+N_2+N_3+N_4} R(t_l) \quad (A.46)$$

It is postulated that the most likely situation in a test of $N_1 + N_2 + N_3 + N_4$ devices is the one described. Consequently, there must be a maximum probability of this event occurring. Therefore, constants s, p and C in the model can be estimated by calculating the values that make the function maximum

$$\ln \Omega = \sum_{i=1}^{N_1} \ln F(t_i) + \sum_{j=N_1+1}^{N_1+N_2} \ln\left(F(t_j) - F(t_j - T_j)\right) + \sum_{k=N_1+N_2+1}^{N_1+N_2+N_3} \ln pdf(t_k) + \sum_{l=N_1+N_2+N_3+1}^{N_1+N_2+N_3+N_4} \ln R(t_l) \quad (A.47)$$

Let $\hat{s}$, $\hat{p}$ and $\hat{C}$ be the estimated values for the constants in the model. The following is therefore verified

$$\Lambda(\hat{s}, \hat{p}, \hat{C}) = \ln \Omega(\hat{s}, \hat{p}, \hat{C}) \quad (A.48)$$

$$\frac{\partial \Lambda}{\partial \hat{s}} = \frac{\partial \Lambda}{\partial \hat{p}} = \frac{\partial \Lambda}{\partial \hat{C}} = 0 \quad (A.49)$$

An estimation of the covariance matrix for the parameters in the model is (the inverse of the Fisher information matrix):

$$\left[\hat{V}_{ij}\right] = \begin{bmatrix} -\dfrac{\partial^2 \Lambda}{\partial \hat{s}^2} & -\dfrac{\partial^2 \Lambda}{\partial \hat{s} \partial \hat{p}} & -\dfrac{\partial^2 \Lambda}{\partial \hat{s} \partial \hat{C}} \\ -\dfrac{\partial^2 \Lambda}{\partial \hat{s} \partial \hat{p}} & -\dfrac{\partial^2 \Lambda}{\partial \hat{p}^2} & -\dfrac{\partial^2 \Lambda}{\partial \hat{p} \partial \hat{C}} \\ -\dfrac{\partial^2 \Lambda}{\partial \hat{s} \partial \hat{C}} & -\dfrac{\partial^2 \Lambda}{\partial \hat{p} \partial \hat{C}} & -\dfrac{\partial^2 \Lambda}{\partial \hat{C}^2} \end{bmatrix}^{-1} \quad (A.50)$$

If the number of devices used to feed the maximum likelihood estimator is large, then random variables $\hat{s}$, $\hat{p}$ and $\hat{C}$ are respectively distributed according to normal distributions $N(s, \hat{V}_{11}^{1/2})$, $N(p, \hat{V}_{22}^{1/2})$ and $N(C, \hat{V}_{33}^{1/2})$ (see Nelson, 2004). Therefore, for a confidence level of 96 per cent, we can write

$$s = \hat{s} \pm 2\hat{V}_{11}^{1/2} \tag{A.51}$$

$$p = \hat{p} \pm 2\hat{V}_{22}^{1/2} \tag{A.52}$$

$$C = \hat{C} \pm 2\hat{V}_{33}^{1/2} \tag{A.53}$$

For an estimation of the parameters (and their errors) to be adequate, the number of failed devices should also be high. For a more in-depth analysis of the maximum likelihood estimator and the calculation of the confidence intervals associated with the constants in some models, see Nelson (2004).

The following matrix expression enables us to calculate the second derivatives in the environment of a point that verifies extreme condition (A.49):

$$\begin{Bmatrix} \frac{\partial^2\Lambda}{\partial\hat{s}^2}\varepsilon_s^2 \\ \frac{\partial^2\Lambda}{\partial\hat{p}^2}\varepsilon_p^2 \\ \frac{\partial^2\Lambda}{\partial\hat{C}^2}\varepsilon_C^2 \\ \frac{\partial^2\Lambda}{\partial\hat{s}\partial\hat{p}}\varepsilon_s\varepsilon_p \\ \frac{\partial^2\Lambda}{\partial\hat{s}\partial\hat{C}}\varepsilon_s\varepsilon_C \\ \frac{\partial^2\Lambda}{\partial\hat{p}\partial\hat{C}}\varepsilon_p\varepsilon_C \end{Bmatrix} = \begin{bmatrix} a & a & a & a & b & b & c & c & c & c & c \\ a & a & a & a & c & c & b & b & c & c & c \\ a & a & a & a & c & c & c & c & b & b & c \\ d & -d & -d & d & 0 & 0 & 0 & 0 & 0 & 0 & 0 \\ d & -d & d & -d & 0 & 0 & 0 & 0 & 0 & 0 & 0 \\ d & d & -d & -d & 0 & 0 & 0 & 0 & 0 & 0 & 0 \end{bmatrix} \begin{Bmatrix} \Lambda(\hat{s}+\varepsilon_s,\hat{p}+\varepsilon_p,\hat{C}+\varepsilon_C) \\ \Lambda(\hat{s}+\varepsilon_s,\hat{p}-\varepsilon_p,\hat{C}-\varepsilon_C) \\ \Lambda(\hat{s}-\varepsilon_s,\hat{p}+\varepsilon_p,\hat{C}-\varepsilon_C) \\ \Lambda(\hat{s}-\varepsilon_s,\hat{p}-\varepsilon_p,\hat{C}+\varepsilon_C) \\ \Lambda(\hat{s}+\alpha\varepsilon_s,\hat{p},\hat{C}) \\ \Lambda(\hat{s}-\alpha\varepsilon_s,\hat{p},\hat{C}) \\ \Lambda(\hat{s},\hat{p}+\alpha\varepsilon_p,\hat{C}) \\ \Lambda(\hat{s},\hat{p}-\alpha\varepsilon_p,\hat{C}) \\ \Lambda(\hat{s},\hat{p},\hat{C}+\alpha\varepsilon_C) \\ \Lambda(\hat{s},\hat{p},\hat{C}-\alpha\varepsilon_C) \\ \Lambda(\hat{s},\hat{p},\hat{C}) \end{Bmatrix}$$

$$\begin{aligned} a &= \frac{\sqrt{11}-2}{15\sqrt{11}-44} = 0.2290,\ b = \frac{9-2\sqrt{11}}{15\sqrt{11}-44} = 0.4117, \\ c &= \frac{-2}{15\sqrt{11}-44} = -0.3479,\ d = 1/4 \\ \alpha &= \sqrt{\sqrt{11}-2} = 1.147 \end{aligned} \tag{A.54}$$

The estimation requires evaluating function (A.48) at 11 different points obtained by choosing $0 < \varepsilon_s \ll s$, $0 < \varepsilon_p \ll |p|$ and $0 < \varepsilon_C \ll C$.[4]

A.8 Design of experiments

As we have seen throughout the book, a designer must: 1) know the transfer functions, 2) characterize the noise affecting the operation and design parameters and the responses, 3) evaluate the quality loss, and 4) know the failure criteria for each failure mode. Sometimes the phenomenon under study is well known, and it is possible to write a functional relationship based on theoretical considerations. However, the mechanisms underlying the process are often not sufficiently understood, or are very complicated, which makes it impossible to postulate an exact model based on theory. Under these circumstances, an empirical or semi-empirical model can be useful, particularly for approximating the response based on only a limited range of the variables. The basic problem facing the design of experiments is deciding what points of the input variables will be better for revealing important aspects of the situation of interest. The question of where the experimental points should be placed is a cyclical problem because if we knew the shape of the response, we could decide where the points should be, but finding out about the response is precisely the objective of the investigation. Fortunately, this circularity is not catastrophic, especially when the experiments can be conducted sequentially, so that the information obtained in a set of experiments can directly influence the choice of the next experimentation points. In this situation, strategy is usually much more important than knowledge because it helps increase the amount of information available per unit of the experimenter's time. Conclusions can easily be drawn from a well-designed experiment, even if the methods of analysis used are quite elementary. On the contrary, the most sophisticated statistical analysis techniques cannot save a series of poorly designed experiments (see Box et al., 1978; Taguchi, 1987).

A.8.1 Regression analysis

The designer seeks the average value of response η depending on a series of variables $x_1, \ldots, x_k$. However, due to experimental error ε, he obtains the following random variable:[5]

$$y = \beta_0 + \beta_1 x_1 + \beta_2 x_2 + \ldots + \beta_k x_k + \varepsilon \tag{A.55}$$

If the designer performs N measurements (using different values for the k variables x_i), he can write the results obtained in matrix form

$$\begin{Bmatrix} y_1 \\ \vdots \\ y_N \end{Bmatrix} = \begin{bmatrix} 1 & x_{11} & \cdots & x_{1k} \\ \vdots & \vdots & \ddots & \vdots \\ 1 & x_{N1} & \cdots & x_{Nk} \end{bmatrix} \begin{Bmatrix} \beta_0 \\ \beta_1 \\ \vdots \\ \beta_k \end{Bmatrix} + \begin{Bmatrix} \varepsilon_1 \\ \vdots \\ \varepsilon_N \end{Bmatrix} \tag{A.56}$$

where each column in the matrix contains the different values that a particular variable has taken over successive tests.[6] This matrix is called experimental matrix X. Because there will be no confusion, we will use the same letter to define the variable $y \in \mathbb{R}$ in equation (A.55) and vector $y^t = (y_1, \ldots, y_N) \in \mathbb{R}^N$ in equation (A.56). If we define vectors $\beta^t = (\beta_0, \beta_1, \ldots, \beta_k) \in \mathbb{R}^{k+1}$ and $\varepsilon^t = (\varepsilon_1, \ldots, \varepsilon_N) \in \mathbb{R}^N$, the above matrix equation is

$$y = X\beta + \varepsilon \tag{A.57}$$

This relationship is unknown because the designer does not know the value of vector β. The objective is to use the design of experiments to replace unknown equation (A.57) with a similar one in which β is approximated by the estimated value $\hat{\beta}$. This approximate equation will provide an estimated result for vector y, which we will call $\hat{y}$. Consequently, the designer replaces (A.57) with

$$\hat{y} = X\hat{\beta} \tag{A.58}$$

The difference between the actual value and the experimental estimation is $\delta = y - \hat{y}$. The way to determine $\hat{\beta}$ is by reducing the error committed at all experimental points to a minimum. This error is

$$\delta^t\delta = (y - \hat{y})^t (y - \hat{y}) = y^t y - y^t X\hat{\beta} - (X\hat{\beta})^t y + \hat{\beta} X^t X\hat{\beta} \tag{A.59}$$

The minimum condition leads to

$$\frac{\partial(\delta^t\delta)}{\partial\hat{\beta}} = -2X^t y + 2X^t X\hat{\beta} = 0 \tag{A.60}$$

From here, we deduce the value of $\hat{\beta}$

$$\hat{\beta} = (X^t X)^{-1} X^t y \tag{A.61}$$

Matrix X^tX is a symmetrical matrix of size $k + 1$, whose elements are the scalar products of the column vectors in the experimental matrix. Therefore, if the column vectors are orthogonal to each other, matrix X^tX is diagonal. In addition, the elements on the diagonal are all positive (different from zero) because they are the square of the modulus of each vector. Consequently, an experimental matrix made up of vectors that are orthogonal to each other leads to an estimation of $\hat{\beta}$ because matrix X^tX is invertible. The orthogonality of the column vectors means that the number of column vectors may not exceed the dimension of the vector. Thus, $N \geq k + 1$ must always be met.

Equations (A.57), (A.58) and (A.61) provide the estimator as a function of the error

$$\hat{\beta} = \beta + (X^tX)^{-1} X^t\varepsilon \tag{A.62}$$

According to the hypothesis stating that the errors are identically and independently distributed with null mean $E[\varepsilon] = 0$ and constant variance $E[\varepsilon\varepsilon^t] = \sigma^2 I$, we will have

$$E[\hat{\beta}] = E[\beta + (X^tX)^{-1} X^t\varepsilon] = E[\beta] + (X^tX)^{-1} X^tE[\varepsilon] = \beta \tag{A.63}$$

$$E[(\hat{\beta} - \beta)(\hat{\beta} - \beta)^t] = E[\hat{\beta}\hat{\beta}^t] - \beta\beta^t \tag{A.64}$$

$$E[\hat{\beta}\hat{\beta}^t] = E[(\beta + (X^tX)^{-1} X^t\varepsilon)(\beta^t + \varepsilon^tX(X^tX)^{-1})] = \beta\beta^t + \sigma^2 (X^tX)^{-1} \tag{A.65}$$

$$E[\hat{y}\hat{y}^t] = E[X(X^tX)^{-1} X^t\varepsilon\varepsilon^tX(X^tX)^{-1} X^t] = \sigma^2X(X^tX)^{-1} X^t \tag{A.66}$$

The last two expressions prove that if the column vectors in the experimental matrix are orthogonal to each other (which is achieved by choosing the appropriate experimental points), the parameter estimators are not correlated to each other. Moreover, when this occurs, the variance associated with the estimators is minimal. Finally, if the errors are distributed normally according to $N(0, \sigma^2)$, the $\hat{\beta}$ estimators are also distributed normally according to $N(\hat{\beta},(X^tX)^{-1}\sigma^2)$.

If we included variables $x_1, \ldots, x_k$ in the model (which we will call block A), and wanted to add other new variables $x_{k+1}, x_{k+2}, \ldots$, (which we will call block B), we would have

$$\beta^t = (\beta_A{}^t, \beta_B{}^t) \tag{A.67}$$

$$\hat{y} = [X_A X_B]\begin{Bmatrix} \hat{\beta}_A \\ \hat{\beta}_B \end{Bmatrix} \tag{A.68}$$

$$\hat{\beta} = \begin{Bmatrix} \hat{\beta}_A \\ \hat{\beta}_B \end{Bmatrix} = \begin{Bmatrix} \beta_A \\ \beta_A \end{Bmatrix} + \left(\begin{bmatrix} X_A{}^t \\ X_B{}^t \end{bmatrix} [X_A X_B] \right)^{-1} \begin{bmatrix} X_A{}^t \\ X_B{}^t \end{bmatrix} \varepsilon \tag{A.69}$$

$$\begin{bmatrix} X_A^t X_A & X_A^t X_B \\ X_B^t X_A & X_B^t X_B \end{bmatrix} \begin{Bmatrix} \hat{\beta}_A - \beta_A \\ \hat{\beta}_B - \beta_B \end{Bmatrix} = \begin{bmatrix} X_A^t \\ X_B^t \end{bmatrix} \varepsilon \tag{A.70}$$

By taking the mathematical expectation in the above matrix expression, we get the following two vector equations:

$$X_A{}^t X_A(E[\hat{\beta}_A] - \beta_A) + X_A{}^t X_B(E[\hat{\beta}_B] - \beta_B) = 0 \quad (A.71)$$

$$X_B{}^t X_A(E[\hat{\beta}_A] - \beta_A) + X_B{}^t X_B(E[\hat{\beta}_B] - \beta_B) = 0 \quad (A.72)$$

These expressions are simplified if the columns in X_A are orthogonal to those in X_B, resulting in

$$E[\hat{\beta}_A] = \beta_A \quad (A.73)$$

$$E[\hat{\beta}_B] = \beta_B \quad (A.74)$$

The independence of the estimators requires that the columns in the experimental matrix be orthogonal.

A.8.2 Two-level factorial designs

In all of the previous chapters, the parameter $m_i \in [\underline{m}_i, \overline{m}_i]$ was defined in terms of the dimensionless variable x_i according to $m_i = \frac{\overline{m}_i + \underline{m}_i}{2} + \frac{\overline{m}_i - \underline{m}_i}{2} x_i$, where $i = 1,2,\ldots,q$. As we will see, the fact that variables x_i take values ±1 at the extremes of interval $[\underline{m}_i, \overline{m}_i]$ is also fundamental for the design of experiments because it naturally leads to the column vectors in the experimental matrix being orthogonal. Although the x_i variables are continuous, the factorial designs that will be presented also work for discrete variables. For example, $x_1 = -1$ can mean 'material A,' and $x_1 = +1$, 'material B.' For this reason, in the context of experimental design, each x_i is indistinctly referred to as a factor or variable.

Factorial designs take the experimental points at the extremes of the intervals, i.e., cause the variables to take the values $x_i = \pm 1$. Because each variable can take two values, the number of tests will be 2^k due to the symmetry between the variables. If we number each test using the variable $j = 0,1,2,3,\ldots,2^k - 1$, each row in the experimental matrix (each test) is defined by the values

$$x_1 = (-1)^{bit_0(j)}, \ldots, x_k = (-1)^{bit_{k-1}(j)} \quad (A.75)$$

This ensures the orthogonality of the experimental matrix in a natural way because there are always as many ones as minus ones in each column. A full factorial design enables us to adjust the following function

$$y = \beta_0 + \sum_{i=1}^{k} \beta_i x_i + \sum_{i=1}^{k} \sum_{\substack{j=1 \\ j \neq i}}^{k} \beta_{ij} x_i x_j + \sum_{i=1}^{k} \sum_{\substack{j=1 \\ j \neq i}}^{k} \sum_{\substack{k=1 \\ k \neq j \\ k \neq i}}^{k} \beta_{ijk} x_i x_j x_k + \ldots \qquad (A.76)$$

Each of the addends in (A.76) is an *effect*.[7] The number of factors that are multiplied by each other is the *order of the effect*. The number of effects whose order is f is $\binom{k}{f}$, where k is the number of variables. The experimental matrix (remember that the elements in the first column are +1) is constructed by generating the k first columns using expressions (A.75), and then the remaining columns by multiplying the previous columns according to the definition of the effect. Thus, the column in the experimental matrix associated with effect $x_1x_2x_5$ is calculated by multiplying the columns in the experimental matrix associated with factors x_1, x_2 and x_5.

The constant in the model associated with an effect can be calculated using the expression

$$\hat{\beta}_{ij\ldots k} = \frac{\sum_{i=1}^{2^k} \left(\frac{x_i x_j \ldots x_k + 1}{2} y_i \right) + \sum_{i=1}^{2^k} \left(\frac{x_i x_j \ldots x_k - 1}{2} y_i \right)}{2^k} \qquad (A.77)$$

This expression calculates the difference between the average of the responses of the effect in question at a high level and the average of the responses with the effect at a low level. The rest of the effects will take a high and low value the same number of times, and will therefore disappear when added together. This procedure offers the following advantages:

1) All of the observations are used to contribute information in every constant.
2) Each effect is determined with the precision of a difference replicated at least the number of times equal to the number of factors.
3) Fewer tests are needed to achieve the same precision as a 'one factor at a time' type method.
4) Because it is an average, the Central Limit Theorem states that the errors associated with the estimation of the constants tend to be distributed by a normal distribution with a lower standard deviation than measured, even if the individual observations on which they are based are not normal. Thus, statistical methods that do not depend directly on distributions of individual observations, but rather on the distribution of one or more observation averages, tend to be insensitive to non-normality.[8]

In expression (A.77), each average contains 2^{k-1} observations, and their variance will be $\sigma^2/2^{k-1}$. The variance of each estimator $\hat{\beta}$ will be $\sigma^2/2^k$. It follows that the experimental error tends to zero as the number of factors increases. An estimation of σ^2 is $\hat{\sigma}^2$ (A.7), which can be obtained by replicating one of the tests n times (for example, the central point). If 2^k is larger enough, the effects can be controlled with a N(0,1) (Box et al., 1978).

A.8.3 Two-level fractional factorial designs

When the number of factors is high, a full factorial design is capable of solving up to very high-order interactions, but such interactions will normally be negligible if the variation range of the factors is small. In that case, these interactions can be confounded with each other, or with the main effects, without losing too much information. Two-level fractional factorial designs are written as 2^{k-r}, where k is the number of factors and 2^{k-r} is the number of tests to be performed.[9]

A full factorial design does not mix any effects. When experiments are eliminated, effects are mixed. The resolution of a design is the sum of the order of the effects that are confounded. A design with resolution R confounds effects of order f with effects of order $R - f$. For example, a resolution III design confounds the order 0 effect with the order 3 effects, and the order 1 effects with the order 2 effects. A resolution V design confounds the order 0 effect with the order 5 effects, the order 1 effects with the order 4 effects, and the order 2 effects with the order 3 effects.[10] The fitting of a hyperplane requires at least resolution III, while the fitting of a quadratic form requires at least resolution V.[11]

A.8.4 Resolution III

The procedure for constructing resolution III designs involves making an effect, whose order is higher than 1, coincide with a new variable. Because the total number of effects with an order greater than one obtained from a set of s variables is $2^s - 1$, these designs can accommodate a number k of variables between s and $2^s - 1$, i.e., $k \in [s, 2^s - 1]$. The number of new factors that the order f effects are capable of accommodating is $s!/(s - f)!/f!$. Table A.1 shows the number of extra factors that a full factorial design with s variables can accommodate, depending on the order of the effect. When all of the effects are used to accommodate new factors, saturated 2_{III}^{k-r} Placket-Burman designs are obtained, which can handle the highest number of variables with the smallest number of tests. For example, a full factorial design with $s = 12$ variables can solve a resolution III hyperplane for $k = 4095$ variables. The maximum number of factors that a complete factorial design with 2^s experiments can accommodate without losing resolution III is

$$k = 2^s - 1 \tag{A.78}$$

A.8.5 Resolution V

The procedure for constructing a resolution V design involves making an order 4 or higher effect coincide with a new variable. The maximum number of variables accommodated by the effects of a given order is limited by the condition that the effects assigned to new variables must have two different variables. If the first order f_i interaction assigned to a new variable is $x_{f_i}x_{f_i-1}...x_1$, the second must be $x_{f_i+2}x_{f_i+1}...x_3$, the third must be $x_{f_i+4}x_{f_i+3}...x_5$, the last one must be $x_{f_i+p}x_{f_i+p-1}...x_{1+p}$. Note that $f_i + p \leq s$, and $2(k_i - 1) = p$ must be verified, where k_i is the number of new factors accommodated by the order f_i effects. From the last two expressions, it follows that $k_i \leq (s - f_i + 2)/2$. If the order of the effect chosen for the first new variables is $f_1 \geq 4$, the order of the effect chosen for the following variables must be $f_2 \geq f_1 + 3$ (this avoids the loss of resolution described in the footnote 10). In general, the order of the effects that accommodate new variables must meet $s \geq f_{j+1} \geq f_j + 3$. For example, with $s = 4$ we can create a resolution V design for $k = 5$ variables if $x_1,..., x_4$ are the first four variables and the fifth variable is $x_5 = x_1x_2x_3x_4$. The maximum number of variables that can be accommodated by a resolution V design starting with $f_1 = 4$ is[12]

$$k = s + \sum_{j=1}^{\mathrm{int}\left(\frac{s-1}{3}\right)} \mathrm{int}\left(\frac{s+1-3j}{2}\right) \tag{A.79}$$

This number appears in the last column of Table A.1.

A.8.6 Fitting of second-order surfaces

With k factors, the minimum number of experiments for adjusting a full second-order surface is $(k + 1)(k + 2)/2$, $k + 1$ experiments for the linear part, k experiments for the quadratic terms of each factor, and $k(k - 1)/2$

Table A.1 Number of variables for resolution III and V fractional factorial designs

s	Order of the effect												$k = 2^s - 1$ R=III	k R=V
	1	**2**	**3**	**4**	**5**	**6**	**7**	**8**	**9**	**10**	**11**	**12**		
1	1	0	0	0	0	0	0	0	0	0	0	0	1	1
2	2	1	0	0	0	0	0	0	0	0	0	0	3	2
3	3	3	1	0	0	0	0	0	0	0	0	0	7	3
4	4	6	4	1	0	0	0	0	0	0	0	0	15	5
5	5	10	10	5	1	0	0	0	0	0	0	0	31	6
6	6	15	20	15	6	1	0	0	0	0	0	0	63	8
7	7	21	35	35	21	7	1	0	0	0	0	0	127	10
8	8	28	56	70	56	28	8	1	0	0	0	0	255	12
9	9	36	84	126	126	84	36	9	1	0	0	0	511	14
10	10	45	120	210	252	210	120	45	10	1	0	0	1023	17
11	11	55	165	330	462	462	330	165	55	11	1	0	2047	19
12	12	66	220	495	792	924	792	495	220	66	12	1	4095	22

experiments for the cross terms. From expression (A.79), we can see that a resolution V fractional factorial design has 2^{s_V} experiments where s_V is the first integer that verifies the inequation

$$s_V + \sum_{j=1}^{\mathrm{int}\left(\frac{s_V-1}{3}\right)} \mathrm{int}\left(\frac{s_V+1-3j}{2}\right) \geq k \tag{A.80}$$

A resolution III fractional factorial design has $2^{s_{III}}$ experiments where

$$s_{III} = k - \mathrm{int}\left(k - \frac{\ln(k+1)}{\ln 2}\right) \tag{A.81}$$

In any factorial design, fractional or otherwise, the quadratic terms of the variables are alloyed with the average value of the response due to the fact that $x_i^2 = +1$. In order to evaluate the quadratic terms, it is necessary to increase the number of factor levels from two to a minimum of three. The simplest way to raise the level of all factors to three is by adding a central point, replicated if possible, on which $x_i = 0$, where $i = 1,\ldots,k$ and k is the number of variables. The central point allows us to test the importance of the quadratic terms, but not solve them.[13] If the quadratic terms are important, a star-shaped design must be added, with the points on the hypersphere, whose center is the central point, and radius α. Each factor, in coded variables, will take the values −α and α. The star is constructed in such a way that when a factor x_i is at a level different from zero, all of the other factors are placed at level zero. Thus, the fractional factorial design with $m = 2^{sv}$ experiments is expanded with a star, which means another $2k$ experiments for the points of the star, and n experiments in the central point, which makes a total of $N = m + 2k + n$ experiments.[14] The number of constants in the model divided by the number of tests in the design is the efficiency of the experimental design

$$\theta = \frac{(k+1)(k+2)}{2N} \tag{A.82}$$

Table A.2 shows the relationship between the number of variables, the number of tests, and the efficiency for resolution III and V designs with a single central point.

The quadratic function that must be adjusted to achieve the orthogonality of the experimental matrix is

$$y = \beta_0 + \sum_i \beta_i x_i + \sum_i \beta_{ii}(x_i^2 - c) + \sum_i \sum_j \beta_{ij} x_i x_j \tag{A.83}$$

Table A.3 outlines the experimental matrix generated by applying the fractional factorial, central points, and points of the star to expression

Table A.2 Efficiency of a resolution V fractional factorial design with a star and a central point as a function of the number of variables

k	1	2	3	4	5	6	7	8	9	10	11	12	13	14	15	16	17	18	19	20
s_{III}	1	2	2	3	3	3	3	4	4	4	4	4	4	4	4	5	5	5	5	5
$2^{s_{III}}$	2	4	4	8	8	8	8	16	16	16	16	16	16	16	16	32	32	32	32	32
s_V	1	2	3	4	4	5	6	6	7	7	8	8	9	9	10	10	10	11	11	12
N	5	9	15	25	27	45	79	81	147	149	279	281	539	541	1055	1057	1059	2085	2087	4137
θ	0.60	0.67	0.67	0.60	0.78	0.62	0.46	0.56	0.37	0.44	0.28	0.32	0.19	0.22	0.13	0.14	0.16	0.09	0.10	0.06

Table A.3 Experimental matrix for a resolution V fractional factorial design with n central points and one star

$m = 2^{S_V}$	1	FRACTIONAL FACTORIAL							1-c	...	...	...	1-c
	...								...				...
	1								1-c	...	...	...	1-c
	1	0	...		...		...	0	-c	...	...	...	-c
n	...	...											...
	1	0	...		...		...	0	-c	...	...	...	-c
	1	α	0	...	0	0	...	0	α^2-c	-c	-c	...	-c
	...	$-\alpha$	0	...	...	...		...	α^2-c	-c	-c	...	-c
		0	α	...	...	...		...	-c	α^2-c	-c	...	-c
$2k$		0	$-\alpha$	...	...	...		...	-c	α^2-c	-c	...	-c
		...	...	...	0	...		...	...				...
	...	...	...	0	α	...		...	...			...	α^2-c
	1	0	...	0	$-\alpha$	0	...	0	-c	...	...	-c	α^2-c
	1	k				$m-k$			k				

(A.83). To maintain the orthogonality of the columns in the experimental matrix, the following must be verified:

$$(1 - c)m + n(-c) + 2(\alpha^2 - c) + (2k - 2)(-c) = 0 \qquad \text{(A.84)}$$

$$m(1 - c)^2 + nc^2 - 4c(\alpha^2 - c) + (2k - 4)c^2 = 0 \qquad \text{(A.85)}$$

These two equations define the values:[15]

$$\alpha = \sqrt{\frac{m}{2}\left(\sqrt{\frac{n}{m}} - 1\right)} \qquad \text{(A.86)}$$

$$c = \sqrt{\frac{m}{N}} \qquad \text{(A.87)}$$

In addition to the second-order design presented here, there are many others with different properties, such as 3^k factorial designs, which are orthogonal designs when the corrected quadratic equation (A.83) is used. However, a drawback of such designs is that the variance of the response surface does generally depend on the direction on which the experimental points are found.

A.8.7 Fitting of first-order surfaces

To calculate only the hyperplane, it is good to know what minimum set of rows in the resolution V design has resolution III. In general, the resolution III design is not completely embedded in the resolution V design. The procedure for generating a resolution V experimental matrix that makes use of the tests in the resolution III design involves writing both matrices and joining them. The duplicate tests cannot be eliminated from the experimental matrix, but they can be removed from the experimental execution.

Table A.4 shows a resolution III design for $k = 5$ factors ($s_{III} = 3$). Table A.5 shows a resolution V design for the same number of factors ($s_V = 4$). Finally, Table A.6 shows a resolution V design that makes use of all the tests performed by the resolution III design. Remember that it is advisable to add a central point to the resolution III fractional factorial design to determine when to execute the resolution V fractional factorial design along with its star.[16] In addition, although we run the 20 tests shown in Table A.6, the orthogonality of the experimental matrix requires duplicating the rows corresponding to the tests that both designs share. For this reason, the star required for evaluating the quadratic terms (Eqs. (A.86) and (A.87)) must be defined with $m = 24$ and $N = 35$.

Table A.4 Experimental design with resolution III for 5 factors

Test	x_1	x_2	x_3	$x_4 = x_1x_2$	$x_5 = x_1x_3$	
1	1	1	1	1	1	V
2	–1	1	1	–1	–1	
3	1	–1	1	–1	1	V
4	–1	–1	1	1	–1	
5	1	1	–1	1	–1	V
6	–1	1	–1	–1	1	
7	1	–1	–1	–1	–1	V
8	–1	–1	–1	1	1	

Note: The last column marks the tests that also appear in the resolution V design

Table A.5 Experimental design with resolution V for 5 factors

Test	x_1	x_2	x_3	x_4	$x_5 = x_1x_2x_3x_4$	
1	1	1	1	1	1	III
2	–1	1	1	1	–1	
3	1	–1	1	1	–1	
4	–1	–1	1	1	1	
5	1	1	–1	1	–1	III
6	–1	1	–1	1	1	
7	1	–1	–1	1	1	
8	–1	–1	–1	1	–1	
9	1	1	1	–1	–1	
10	–1	1	1	–1	1	
11	1	–1	1	–1	1	III
12	–1	–1	1	–1	–1	
13	1	1	–1	–1	1	
14	–1	1	–1	–1	–1	
15	1	–1	–1	–1	–1	III
16	–1	–1	–1	–1	1	

Note: The last column marks the tests that also appear in the resolution III design

Table A.6 **Experimental design with resolution III+V for 5 factors**

Test	x_1	x_2	x_3	x_4	x_5	
1	1	1	1	1	1	III+V
2	−1	1	1	1	−1	V
3	1	−1	1	1	−1	V
4	−1	−1	1	1	1	V
5	1	1	−1	1	−1	III+V
6	−1	1	−1	1	1	V
7	1	−1	−1	1	1	V
8	−1	−1	−1	1	−1	V
9	1	1	1	−1	−1	V
10	−1	1	1	−1	1	V
11	1	−1	1	−1	1	III+V
12	−1	−1	1	−1	−1	V
13	1	1	−1	−1	1	V
14	−1	1	−1	−1	−1	V
15	1	−1	−1	−1	−1	III+V
16	−1	−1	−1	−1	1	V
17	−1	1	1	−1	−1	III
18	−1	−1	1	1	−1	III
19	−1	1	−1	−1	1	III
20	−1	−1	−1	1	1	III

Note: The last column marks the origin of the tests.

If we do not replicate the duplicated tests (marked with III+V in Table A.6), the efficiency of the III+V design is 68 per cent, slightly lower than 78 per cent, which is the efficiency of a pure resolution V design. However, this decrease in the efficiency is widely compensated if a large number of steps is required to reach the point where the quadratic terms are significant.

A.8.8 Optimization

The objective is to find an extreme of the y function.[17] The tangent hyperplane is calculated from the resolution III fractional factorial using the following expression (the central point is not used, except to highlight the importance of the non-linear terms):

$$\begin{bmatrix} \hat{\beta}_0 \\ \hat{\beta} \end{bmatrix} = 2^{-s_{III}} X_{III}^t y \tag{A.88}$$

where X_{III} is the experimental matrix of the resolution III fractional factorial design. (Note that, for convenience in later calculations, the constant term has been separated from the hyperplane.)

The hyperplane is written:

$$\hat{y} = \hat{\beta}_0 + \hat{\beta}^t x \tag{A.89}$$

The normal to the hyperplane is (pointing in the ascending direction of y):

$$n^t = (1, -\hat{\beta}^t) \tag{A.90}$$

The direction for finding a maximum value of $\hat{y}$ is given by ($\lambda > 0$):

$$x = -\lambda \hat{\beta} \tag{A.91}$$

Because x is a vector defined by coded variables, it is not advisable to venture outside of the explored area. By imposing $x^t x = 1$, we find that a point closest to the maximum (changing the sign will bring it closer to the minimum) is

$$x = -\frac{\hat{\beta}}{\sqrt{\hat{\beta}^t \hat{\beta}}} \tag{A.92}$$

If the quadratic terms in $x = 0$ are not very significant, the exploration length can be doubled, tripled, etc. At the new point $x = -i\hat{\beta} / \sqrt{\hat{\beta}^t \hat{\beta}}$, where $i = 1,2,\ldots$, a central point and resolution III fractional factorial will again be evaluated. Near the extreme, $\sqrt{\hat{\beta}^t \hat{\beta}} \ll 1$ is verified, which helps detect the proximity of the extreme. This can be corroborated by comparing $\hat{\beta}_0$ with the test result at the central point. When the quadratic terms are found to be important, the rest of the experiments needed to increase the resolution to V are conducted. The calculation of the paraboloid near the extreme is performed based on the resolution V fractional factorial plus the tests coming from the resolution III design, the star, and the central point. This design is

$$\begin{bmatrix} \hat{\beta}_0 \\ \hat{\beta} \\ \hat{y} \end{bmatrix} = \left(X_{V+III}^t X_{V+III} \right)^{-1} X_{V+III}^t y \tag{A.93}$$

By reordering and scaling[18] the components of vector γ, we construct matrix $[\gamma]$ and the paraboloid

$$\hat{y} = \hat{\beta}_0 + \hat{\beta}^t x + x^t [\hat{y}] x \tag{A.94}$$

The extreme will be the point that verifies

$$x = -\frac{1}{2}[\hat{y}]^{-1}\hat{\beta}^t \tag{A.95}$$

A.9 Notes

1. For this reason, if the sensitivities of the transfer function are fixed, the greater the number of parameters, the higher the information content and the higher the quality loss. See Chapters 3 and 4.
2. For variable z, distributed according to N(0,1), calculating $z_{\alpha/2}$ means solving the equation $1-\alpha/2 = \int_{-\infty}^{z_{\alpha/2}} (e^{-z^2/2}/\sqrt{2\pi})dz$. For α = 0.04, the confidence coefficient is $1 - \alpha = 0.96$ with the value $z_{\alpha/2} \simeq 2$.
3. There are other models for variable loads, such as the cumulative exposure model described by Nelson (2004). Balakrishnan (2009), Balakrishnan, Xie and Kundu (2009), Balakrishnan and Han (2009), Balakrishnan, Zhang and Xie (2009), Balakrishnan et al. (2007), Kateri and Balakrishnan (2008), Han et al. (2006), Nelson (2004), Gouno, Sen and Balakrishnan (2004) and Khamis and Higgins (1998) have presented an extensive body of work on step-stress models.
4. Estimation uses more points than unknown quantities, and an orthogonal matrix for reducing the error.
5. Although the model is linear in variables, $x_1, \ldots, x_k$, it does not necessarily have to be linear in the original variables. For example, it could be $x_3 = x_1x_2$, $x_3 = \log x_1$, $x_3 = x_2^2$ or, in general, $x_3 = f(x_1, x_2, z)$. This occurs when a variable in the model, for example, x_k, is a function of the previous ones. A typical example in design of experiments is $x_k = x_1x_2 \ldots x_{k-1}$. In general, the number q of parameters in the transfer function does not necessarily have to coincide with the number k of variables in the experimental model.
6. Note that the constant parameter in the model (first column in the experimental matrix) is associated with the fictitious variable x_0, which takes the value +1 in all tests.
7. The linear terms are usually called the main effects, and the non-linear terms that the factorial design is capable of estimating are usually referred to as interaction effects. Thus, interactions are the products of factors that are different from each other, excluding quadratic and cubic terms, etc. In a factorial design with two levels, there can be no variables raised to a power ($x_i^n = 1$ if n is even, and $x_i^n = -1$ if n is odd).
8. The replicas must be genuine; in other words, all steps leading to the conclusion of the experiment must be repeated in full, and not by taking several measurements from the same test (without varying the factors). That would only include the variance due to the measurement.

9. The simplest design of this type takes the highest-order interaction in a full factorial design for $k-1$ factors and confounds it with the factor not included in the full factorial design: $x_k = x_1x_2...x_{k-1}$. Such a design is written as 2^{k-1}, and only has half of the experiments. This design can be used to block an indeterminate factor, such as time, or different lots of material.
10. A resolution V design with five variables verifies $x_1x_2x_3x_4x_5 = 1$. If resolution III is also imposed through the relationship $x_1x_2x_3 = 1$, we immediately discover that the design is resolution II because from the two previous conditions, it follows that $x_4x_5 = 1$.
11. If additional information is available, for example, if the order 1 factors are null, then a resolution III design is sufficient for adjusting the order 2 effects. This is the case of the design used in expression (A.54).
12. The maximum number of new factors that can be accommodated by the effects of order f_j is $k_j = \text{int}(s - f_j + 2)/2$. In addition, $f_j = f_1 + 3(j-1) = 3j + 1 \leq s$ holds.
13. This is obtained by comparing the average of factorial design $\bar{y}_f$ with the average in the center of design $\bar{y}_c$, so that $\bar{y}_f - \bar{y}_c$ is an estimation of the overall surface curvature.
14. Performing more than one experiment on the central point enables us to estimate the standard deviation associated with the experimental errors. It is also useful to make use of the previous tests that have resolution III. In this case, the number of points in the fractional factorial design would be $m = 2^{s_V} + 2^{s_{III}}$.
15. In an experimental design without the fractional factorial, $m = 0$, so that $\alpha = c = 0$. It is therefore impossible to create a star design that is orthogonal. The orthogonality of the experimental design matrix makes it necessary to use the fractional factorial plus the star.
16. Note that adding central points to a factorial design does not break the orthogonality of the columns in the experimental matrix.
17. This procedure is especially useful for decreasing the quality loss of a design when the transfer function is unknown or extraordinarily complicated. To do so, simply replace the y function with the quality loss function throughout the procedure. Papalambros and Wilde (1988) describes techniques for design optimization.
18. The terms outside of the diagonal in $[\gamma]$ are half of the corresponding component of vector γ.

References

Akiyama, K. 1991. *Function Analysis: Systematic Improvement of Quality and Performance*. Cambridge, MA: Productivity Press Inc.

Arcidiacono, G., Capitani, R., Citti, P., Panichi, C. and Rosti, D. 2006. 'A new integrated approach to the design of a race car suspension', Proceedings of ICAD2006, Fourth International Conference on Axiomatic Design, Firence (Italy).

Ba-abbad, M., Nikolaidis, E. and Kapania, R. 2006. 'New approach for system reliability based design optimization', *AIAA journal*, vol. 44, no. 5, pp. 1087–96.

Balakrishnan, N. 2009. 'A synthesis of exact inferential results for exponential step-stress models and associated optimal accelerated life-tests', *Metrika*, vol. 69, pp. 351–96.

Balakrishnan, N. and Han, D. 2009. 'Optimal step-stress testing for progressively type-I censored data from exponential distribution', *Journal of Statistical Planning and Inference*, vol. 139, pp. 1782–98.

Balakrishnan, N., Kundu, D., Ng, H. K. T. and Kannan, N. 2007. 'Point and interval estimation for a simple step-stress model with type-II censoring', *Journal of Quality Technology*, vol. 39, pp. 35–47.

Balakrishnan, N., Xie, Q. and Kundu, D. 2009. 'Exact inference for a simple step-stress model from the exponential distribution under time constraint', *Annals of the Institute of Statistical Mathematics*, vol. 61, pp. 251–74.

Balakrishnan, N., Zhang, L. and Xie, Q. 2009. 'Inference for a simple stepstress model with type-I censoring lognormally distributed lifetimes', *Communications in Statistics – Theory and Methods*, vol. 38, no. 10, pp. 1709–1960.

Bandecchi, M. 2001. 'The ESA Concurrent Design Facility (CDF): concurrent engineering applied to space mission assessments', 2nd Nordic Systems Engineering Boat Seminar, Massimo Bandecchi, Head of Concurrent Design Unit, ESA/ESTEC Noordwijk-NL FinSE2001, 25–26 September 2001.

Bandecchi, M., Melton, B. and Ongaro, F. 1999. 'Concurrent Engineering Applied to Space Mission Assessment and Design', ESA Bulletin No. 99, September 1999.

Benavides, E. M. 2004. 'Reliability and Taguchi's Metric', Lecture, 4th International Conference on Advanced Engineering Design, Glasgow 2004.

Benavides, E. M. 2006. 'Using Quality Loss Function for Preliminary Selection of a Roller Bearing at Early Design Stages', Lecture, 5th International Conference on Advanced Engineering Design, Praga 2006.

Benavides, E. M. 2010. 'Reliability model for step-stress and variable-stress situations', *IEEE Transactions on Reliability*, vol. 60, issue 1, pp. 219–33.

Benavides, E. M. 2011. 'Axiomatic-design approach to the main-bearing configuration of a jet engine with several shafts', Proceedings of ICAD2011,

The Sixth International Conference on Axiomatic Design, Daejeon, 30–31 March 2011.

Benavides, E. M. and García-Rodríguez, L. 2011. 'Extended algorithm for design-matrix reorganization', Proceedings of ICAD2011, The Sixth International Conference on Axiomatic Design, Daejeon, 30–31 March 2011.

Blish, R. C. 1997. 'Temperature Cycling and Thermal Shock Failure Rate Modeling', Proceedings in IEEE International Reliability Physics Symposium, Denver, CO, 8–10 April 1997, pp. 110–17.

Box, G. E. P., Hunter, W. G. and Hunter, J. S. 1978. *Statistics for experimenters: an introduction to design, data analysis, and model building*. New York, NY: John Wiley and Sons.

Brown, C. A. 2005. 'Teaching Axiomatic Design to Engineers – Theory, Applications, and Software', *Journal of Manufacturing Systems*, vol. 24, no. 3.

Cai, C-L., Xiao, R-B. and Yang, P. 2009. 'The method for analyzing and disposing of functional interaction in axiomatic design', *Journal of Mechanical Engineering Science*, Proc. IMechE vol. 224, Part C.

Cebi, S. and Kahraman, C. 2008. 'Axiomatic Design principles under fuzzy environment', Proceedings of 2008 3rd International Conference on Intelligent System and Knowledge Engineering, Xiamen, 17–19 November 2008.

Conçalves-Coelho, A. M. and Mourâo, A. J. F. 2007. 'Axiomatic design as support for decision-making in a design for manufacturing context: a case study', *International Journal of Production Economics*, vol. 109, pp. 81–89.

Cover, T. H. and Thomas, J. A. 2006. *Elements of information theory*, 2nd edn. New Jersey, NJ: John Wiley and Sons.

Cruse, T. A. (ed.) 1997. *Reliability-based mechanical design*. New York, NY: Marcel Dekker, Inc.

Cruse, T. A. and Sankaran, M. 1997. 'Fatigue reliability of gas turbine engine structures', NASA CR-97-206215.

Dean, E. B. 1992. 'Quality Function Deployment for Large Systems', Technical Report: conf-iemc-92.

Dean, E. B. 1993a. 'Genopersistating the System', presented at the AIAA Aerospace Design Conference, Irvine, CA, 16–19 February 1993, AIAA-93-1031.

Dean, E. B. 1993b. 'Why Does It Cost How Much', AIAA Aircraft Design, Systems, and Operations Conference, Monterey, CA, 11–13 August 1993, AIAA-93-3966.

Dean, E. B. 1995. 'Parametric Cost Deployment', Technical Report: NASA-qfd-95-p27, NASA Langley Technical Report Server.

Dean, E. B. 1997. *http://www.go2hill.com/ProjectManagement/FUNCTION.htm*.

Dean, E. B. and Unal, R. 1992. 'Elements of Designing for Cost', AIAA Aerospace Design Conference, Irvine CA, 3–6 February 1992, AIAA-92-1057.

Du, X. and Sudjianto, A. 2004. 'First order saddlepoint approximation for reliability analisis', *AIAA journal*, vol. 42, no. 6, pp. 1199–1207.

El-Haik, B. 2005. *Axiomatic Quality: Integrating Axiomatic Design with Six-Sigma, Reliability, and Quality Engineering*. New Jersey, NJ: John Wiley and Sons.

El-Haik, B. and Yang, K. 1999. 'The components of Complexity in Engineering Design', *IIE Transactions*, vol. 31, issue 10, pp. 925–34.

Fowlkes, W. Y. and Creveling, C. M. 1995. *Engineering Methods for Robust Product Design: Using Taguchi Methods in Technology and Product Development*. London: Addison Wesley.

Fukuda, M. 1995. *Reliability and Degradation of Semiconductor Lasers and LEDs*. Boston, MA: Artech House.

Gebala, D. A., and Suh, N. P. 1992. 'An Application of Axiomatic Design', *Research in Engineering Design*, vol. 3, pp. 149–62.

Glegg, G. L. 1960. *Design of Design*. New York, NY: Cambridge University Press.

Gouno, E., Sen, A. and Balakrishnan, N. 2004. 'Optimal step-stress test under progressive type-I censoring', *IEEE Trans. Reliability*, vol. 53, no. 3, pp. 388–93.

Grandhi, R. and Wang, L. 1999. 'Structural reliability analisis and optimization: Use of approximations', NASA CR-1999-209154.

Han, D., Balakrishnan, N., Sen, A. and Gouno, E. 2006. 'Corrections on Optimal Step-Stress Test under Progressive Type-I Censoring', *IEEE Trans. Reliability*, vol. 55, no. 4, pp. 613–14.

Harris, T. A. 2001. *Rolling Bearing Analysis*, 4th edn. New Jersey, NJ: John Wiley & Sons, p. 1086.

Harrisberger, L. 1966. *Engineermanship: a Philosophy of Design*. Belmont, CA: Brooks/Cole.

Hill, P. H. 1970. *The science of Engineering Design*. New York, NY: Holt, Rinehart and Winston.

Hirani, H. and Suh, N. P. 2005. 'Journal Bearing design using multiobjective genetic algorithm and axiomatic design approaches', *Tribology International*, vol. 38, issue 5, pp. 481–91.

Huang, G. Q. and Jiang, Z. 2002. 'Web-base design review of fuel pumps using fuzzy set theory', *Engineering Applications of Artificial Intelligence*, vol. 15, issue 6, pp. 529–39.

Hubka, V. and Eder, W. E. 1997. 'A scientific approach to Engineering Design,' *Design Studies*, vol. 8, issue 3, pp. 123–37.

Hundal, M. S. 1997. *Systematic Mechanical Design: a cost and management perspective*. New York, NY: ASME Press.

Hyvärinen, L. P. 1968. 'Information theory for systems enginers', in *Lecture notes in operation research and mathematical economics*. Berlin: Springer-Verlag.

IMechE. 2009. *Injection systems for IC engines*. Oxford: Chandos Publishing.

Jacobson, I., Christerson, M., Jonsson, P. and Övergaard, G. 1992. *Object-Oriented Software Engineering: a use case driven approach*. London: Addison Wesley.

Jagannathan, V., Cleetus, K. J., Kannan, R., Matsumoto, A. S. and Lewis, J. W. 1991. 'Computer Support for Concurrent Engineering', *Concurrent Engineering*, vol. 1, no. 5, pp. 14–30.

Jones, D. S. 1989. *Elementary information theory*. Oxford: Oxford University Press.

Kar, A. K. 2000. 'Linking axiomatic design and Taguchi methods via information content in design', Proceedings of ICAD2000, First International Conference on Axiomatic Design, Cambridge, MA, 21–23 June 2000.

Kateri, M. and Balakrishnan, N. 2008. 'Inference for a simple step-stress model with type-II censoring, andWeibull distributed lifetimes', *IEEE Trans. Reliability*, vol. 57, no. 4, pp. 616–26.

Khamis, I. H. and Higgins, J. J. 1998. 'A new model for step-stress testing', *IEEE Trans. Reliability*, vol. 47, no. 2, pp. 131–34.

Kulak, O., Cebi, S. and Kahraman, C. 2010. 'Applications of axiomatic design principles: A literature review', *Expert Systems with Applications*, vol. 37, issue 9, pp. 6705–17.

Kullback, S. 1968. *Information theory and statistics*. New York, NY: Dover Publication.

Lee, T. 2003. *Complexity Theory in Axiomatic Design*, Ph.D. Thesis, Massachusetts Institute of Technology, May 2003.

Lee, T. 2006. 'Optimal Strategy for eliminating coupling terms from a design matrix', *Journal of Integrated Design and Process Technology*, vol. 10, issue 2.

Lee, T. and Jeziorek, N. P. 2006. 'Understanding the value of eliminating an off-diagonal term in a design matrix', Proceedings of ICAD 2006, 4th International Conference on Axiomatic Design, Firenze, 13–16 June 2006.

Levada, S., Meneghini, M., Meneghesso, G. and Zanoni, E. 2005. 'Analysis of DC Current Accelerated Life Tests of GaN LEDs Using a Weibull-Based Statistical Model,' *IEEE Trans. Dev. Mat. Rel.*, vol. 5, no. 4, pp. 688–93.

Lian, Y. and Kim, N-H. 2006. 'Reliability-based design optimization of a transonic compressor', *AIAA journal,* vol. 44, no. 2, pp. 368–75.

Logothetis, N. 1992. *Managing for Total Quality: From Deming to Taguchi and SPC*. London: Prentice Hall.

Matt, D. T. 2007. 'Reducing the structural complexity of growing organizational systems by means of axiomatic designed networks of core competence cells', *Journal of Manufacturing Systems*, vol. 26, no. 3+4, pp. 178–87.

Matt, D. T. 2009. 'Reducing the time dependent complexity in organizational systems using the concept of functional periodicity', Proceeding of ICAD2009, The fifth International Conference on Axiomatic Design, Campus de Caparica, 25–27 March 2009.

Nelson, W. B. 2004. *Accelerated Testing, Statistical Models, Test Plans and Data Analysis*. New York, NY: John Wiley & Sons.

Orloff, M. A. 2006. *Inventive Thinking Through TRIZ: a prectical guide*, 2nd edn. Berlin: Springer.

Ostrofsky, B. 1977. *Design, Planning and Development Methodology*. Englewood Cliffs, NJ: Prentice-Hall.

Papalambros, P. Y. and Wilde, D. J. 1988. *Principles of optimal design: modeling and computation*. Cambridge: Cambridge University Press.

Rackwitz, R. 2001. 'Reliability analysis – a review and some perspectives', *Structural Safety*, vol. 23, issue 4, pp. 365–95.

Rackwitz, R., and Fiessler, B. 1978. 'Structural Reliability under combined random load sequences', *Computers and Structures*, vol. 9, issue 5, pp. 489–94.

Rinderle, J. R., Suh, N. P. 1982. 'Measures of functional coupling in design', *Journal of Engineering for Industry*, vol. 104, no. 4, pp. 383–88.

Rodríguez-Pastor, J. B. and Benavides, E. M. 2011. 'Axiomatic Design of an Airport Passenger Terminal', Proceedings of ICAD2011, The Sixth International Conference on Axiomatic Design, Daejeon, 30–31 March 2011.

Ross, P. J. 1988. *Taguchi Techniques for Quality Engineering: loss function, orthogonal experiments, parameter and tolerance design*. New York, NY: McGraw-Hill, p. 279.

Santos, A. G., Silva, A. and Gonçalves-Coelho, A. 2009. 'The minimum constraint design and the first axiom', Proceeding of ICAD2009, The fifth International Conference on Axiomatic Design, Campus de Caparica, 25–27 March 2009.

Shannon, C. E. 1948. 'A mathematical theory of communication', *Bell System Tech. J.*, vol. 27, pp. 379–423, 623–56.

Shillito, M. L. 1994. *Advanced QFD: linking technology to market and company needs*. New York, NY: John Wiley and Sons.

Shirwaiker, R. A. and Okudan, G. E. 2006. 'TRIZ and Axiomatic Design: a review of manufacturing case-studies and their compatibility', Proceedings of PICMET, Istanbul, 9–13 July 2006.

Simon, H. 1969. *The Science of the Artificial*. Cambridge, MA: MIT Press.

Su, J. C-Y., Chen, S-J. and Lin, Li. 2003. 'A structured approach to measuring functional dependency and sequencing of coupled tasks in engineering design', *Computers & Industrial Engineering*, vol. 45, issue 1, pp. 195–214.

Suh, N. P. 1984. 'Development of the Science Base for the Manufacturing Field Through the Axiomatic Aproach', *Robotics and Computer Integrated Manufacturing*, vol. 1 (3/4), pp. 397–415.

Suh, N. P. 1990. *The principles of Design*. New York, NY: Oxford University Press.

Suh, N. P. 1995. 'Designing-in of quality through Axiomatic Design', *IEEE Transactions on Reliability*, vol. 44, no. 2.

Suh, N. P. 1996. 'Microcellular Plastics', in J. S. Molding (ed.) *Innovation in Polymer Processing*. New York, NY: SPE Books of Hanser Publishers.

Suh, N. P. 1999. 'A theory of complexity, periodicity and the design axioms', *Research in Engineering Design*, vol. 11, no. 2, pp. 116–31.

Suh, N. P. 2001. *Axiomatic Design: advances and applications*. New York, NY: Oxford University Press.

Suh, N. P. 2005. *Complexity: theory and applications*. New York, NY: Oxford University Press.

Suh, N. P. 2007. 'Ergonomics, axiomatic design and complexity theory', *Theoretical Issues in Ergonomics Science*, vol. 8, no. 2, pp. 101–21.

Suh, N. P., Bell, A. C. and Gossard, D. C. 1978. 'On an Axiomatic Approach to Manufacturing Systems', *Journal of Engineering for Industry*, vol. 100, issue 2, pp. 127–30.

Taguchi, G. 1987. *System of experimental design: Engineering methods to optimize quality and minimize costs*, Volumes one and two. White Plains, NY: UNIPUB/Kraus International Publications.

Taguchi, G., Chowdhury, S. and Wu, Y. 2001. *The Mahalanobis-Taguchi system*. New York, NY: McGraw Hill, p. 190.

Taguchi, G., Elsayed, E. A. and Hsiang, T. 1989. *Quality Engineering in Production Systems*. New York, NY: McGraw-Hill, p. 173.

Taguchi, G., Konishi, S. and Wu, Y. 1991. *Taguchi methods: signal-to-noise ratio for quality evaluation*. Dearborn, MI: American Supplier Institute, p. 370.

Taguchi, G., Konishi, S. and Wu, Y. 1992. *Taguchi methods: research and development*. Dearborn, MI: American Supplier Institute, p. 351.

Taguchi, G. and Rajesh, J. 2002. *The Mahalanobis-Taguchi strategy: a pattern technology system*. New York, NY: John Wiley & Sons, p. 234.

Taguchi, G. and Wu, Y. 1989. *Taguchi methods: case studies from the U.S. and Europe*. Dearborn, MI: American Supplier Institute, p. 368.

Tchidi, M. F. and He, Z. 2010. 'Six Sigma Axiomatic Design Method Based on Quality Function Deployment', The 2nd IEEE International Conference on Information Management and Engineering, Chengdu, 16–18 April 2010.

Thompson, G. 1999. *Improving maintainability and reliability through design*. London: Professional Engineering Publishing, p. 216.

Thompson, M. K. 2009. 'Teaching Axiomatic Design in the freshman year: a case study at KAIST', Proceeding of ICAD2009, The fifth International Conference on Axiomatic Design, Campus de Caparica, 25–27 March 2009.

Thompson, M. K., Kwon, O. H. and Park, M. J. 2009. 'The application of Axiomatic Design Theory and Conflict Techniques for the design of intersections: Part 1', Proceeding of ICAD2009, The fifth International Conference on Axiomatic Design, Campus de Caparica, 25–27 March 2009.

Thompson, M. K., Park, M. J., Kwon, O. H., Ibragimova, E., Lee, H. and Myung, S. 2009. 'The application of Axiomatic Design Theory and Conflict Techniques for the design of intersections: Part 2', Proceeding of ICAD2009, The fifth International Conference on Axiomatic Design, Campus de Caparica, 25–27 March 2009.

Thompson, M. K., Thomas, B. C. and Hopkins, J. B. 2009. 'Applying Axiomatic Design to the Educational Process', Proceeding of ICAD2009, The fifth International Conference on Axiomatic Design, Campus de Caparica, 25–27 March 2009.

Tobias, P. A. and Trindade, D. C. 1986. *Applied Reliability*. New York, NY: Van Nostrand Reinhold Company Inc.

Wu, Y. and Wu, A. 2000. *Taguchi Methods for robust design*. New York, NY: The American Society of Mechanical Engineers, p. 334.

Yihai, H., Zhao, M. and Wenbing, C. 2009. 'A technical Framework of the Taguchi System Design Method Base don Axiomatic Design and TRIZ', Proceedings of the 2009 IEEE International Conference on Industrial Engineering and Engineering Management, Hong Kong, 8–11 December 2009.

Yoshikawa, H. 1985. 'Extended General Design Theory', in H. Yoshikawa, *Design Theory for CAD: Proceedings of the IFIP W.G. 5.2 Working Conference on Design Theory for Cad*. Amsterdam: Elsevier.

Youn, J. R., and Suh, N. P., "Processing of microcellular polyester composites", Polymer Composites, Vol. 6, pp. 175–180, 1985.

Zambrano, J. A. A. 2009. *Ampliación del modelo de Diseño Axiomático para el desarrollo de productos con equipos multidisciplinares*, Ph.D. Thesis, Universidad Politécnica de Valencia.

Index

FSC
www.fsc.org
MIX
Papier | Fördert
gute Waldnutzung
FSC® C083411

Kolibri 360 GmbH
Ferdinand-Jühlke-Straße 7
99095 Erfurt, Deutschland
produktsicherheit@kolibri360.de